P9-DUZ-238

SPACE IN THE 21ST CENTURY

SPACE
IN THE
21ST
CENTURY

RICHARD S. LEWIS

COLUMBIA UNIVERSITY PRESS   NEW YORK

Title page art: Artist's concept of the Space Shuttle approaching the United States *Freedom* Space Station, in orbit, circa 1999.

COLUMBIA UNIVERSITY PRESS
NEW YORK OXFORD

Library of Congress Cataloging-in-Publication Data

Lewis, Richard S., 1916–
    Space in the 21st century / Richard S. Lewis.
       p.   cm.
    Includes bibliographical references.
    ISBN 0-231-06304-0 (alk. paper)
    1. Outer space—Exploration. 2. Space colonies. I. Title.
II. Title: Space in the twenty-first century.
    TL790.L395   1990
    919.9′04—dc20   89-22317
                                        CIP

Casebound editions of Columbia University Press books are Smyth-sewn and printed on permanent and durable acid-free paper

Printed in the United States of America

Design by Charles B. Hames

c 10 9 8 7 6 5 4 3 2 1

CONTENTS

1. Beyond Tranquility   1

2. Birth of a Station   19

3. The Grand Design   45

4. Transition   74

5. The New World   99

6. A Manifest Destiny   119

7. Power from the Moon   140

8. A Natural History of Mars   160

9. The Emergence of Mars   188

10. CASTLES In Orbit   206

Epilogue   223

Notes   225

Index   229

SPACE IN THE
21ST CENTURY

O N E

BEYOND

TRANQUILITY

F ROM the perspective of two decades, the landing of *Apollo 11* in the Mare Tranquillitatis was the first step toward human expansion into the Solar System. It was an event of evolutionary scale, portending an environmental transition for the human species on the order of the emergence of life from the sea.

Although *Apollo 11* was popularly regarded as an American victory in the cold war, the space community perceived the development of the Saturn 5–Apollo transportation system as the true beginning of interplanetary travel. For the system was also capable of reaching Mars. Responding to this perception, the Nixon White House appointed a blue ribbon committee, the Space Task Group, and charged it with the mission of devising a post-Apollo program for the exploration of the Solar System. By September 1969, hardly two months after the *Eagle* had landed July 20, a detailed blueprint for the exploration of the Moon and beyond was presented by the National Aeronautics and Space Administration to the Space Task Group. Although never fully realized, this plan, "America's Next Decades in Space," has remained the basic program of space exploration by the United States. When congressional and other critics say that NASA has no plan, they refer simply to the space agency's failure to follow it in an orderly fashion.

"The next natural step for us to take in space," the 1969 plan stated, "will be for us to create permanent space stations in Earth and in lunar orbits with low cost access by reusable chemical and nuclear rocket transportation systems and to utilize these systems in assembling our capability to explore the planet Mars with men, thereby initiating man's permanent occupancy of outer space." Two decades later, the national intention of expanding human presence and activity beyond Earth orbit into the Solar System was proclaimed formally as a goal on February 11, 1988, by President Ronald Reagan. The policy statement was not accompanied by any details indicating how it could be implemented.

In 1969 the logistical strategy for manned exploration of the Solar System was to employ only a few major vehicle systems for a variety of functions. Their elements would be interchangeable and reusable. This development strategy would apply to unmanned as well as manned vehicle systems. It would distinguish future American development in manned space flight from the Soviet practice of constructing nonreusable, noninterchangeable flight systems. Thus, under the program projected by the Space Task Group in 1969, a habitat module developed for a space station in Earth or in lunar orbit could also be used as a crew or cargo module on a trip to Mars.* The space station would be supplied by a fully reusable shuttle. A reusable space "tug" based at the space station in low Earth orbit would provide the means of moving cargo from one orbit to another (e.g., communications satellites from the space station orbit at 250 miles altitude to geostationary orbit at 22,300 miles altitude).

The 1969 vehicle plan included a nuclear-powered shuttle plying between stations in Earth and lunar orbits. The tug could operate as a lunar

---

*Members of President Nixon's Space Task Group were Thomas O. Paine, NASA administrator; Robert C. Seamans, Secretary of the Air Force and former deputy administrator of NASA; Lee A. DuBridge, the president's science adviser, and Vice-President Spiro T. Agnew, in his capacity as chairman of the National Aeronautics and Space Council, adviser to NASA.

---

landing vehicle, depositing explorers and a space station module that would serve as a shelter for them on the surface of the Moon. Clusters of space tug propulsion units could be assembled in low Earth orbit to boost an array of space station modules, with crew and supplies, on an eight-month journey around the sun to Mars.

The blueprint that NASA prepared for the Nixon Space Task Group was drawn from projections of lunar and planetary expeditions by NASA centers and a coterie of academic and private consultants. Although the technology required for these scenarios lay within the capability of American science and engineering, their implementation depended on the socioeconomic and political states of the nation. The development of space flight in America was motivated largely by intraspecific competition, expressed in the 1960s when manned orbital flight began, by the Cold War. It motivated the decision to be first on the Moon. In a more settled atmosphere, it has continued to stimulate a sense of urgency to land on Mars before the Russians do. "America's Next Decades in Space," our first long-range space program was the byproduct of this competitive process.

THE NEXT LOGICAL STEP

Since manned spaceflight began in America with Project Mercury in 1961, the construction of a space station in low Earth orbit has been considered the next logical step by the space community. Looking down on Earth from an altitude of several hundred miles in space conferred upon human beings a new sense of understanding and control of their environment. Here was an outpost of the cosmic frontier, a jumping-off point for the beyond.

The idea was broached in 1869 in an article written by Edward Everett Hale and published by the *Atlantic Monthly*. Entitled "The Brick Moon," the article described a platform 192 feet in diameter that circled the Earth at an altitude of 3600 miles. It served as a navigation beacon for ships at sea and carried a crew of 37 men. A more sophisticated concept was proposed in 1923 by Konstantin Tsiolkovsky, the Russian high school teacher who devised the theoretical foundation for space flight. He described a satellite observatory in Earth orbit. The same year, Hermann Oberth, a Romanian engineer working in Germany, proposed a manned satellite that would provide a platform for Earth observations and also serve as a refueling depot for space vehicles. In 1952 Wernher von Braun described a wheel-shaped platform 250 feet in diameter while he was working for the Army Ballistic Missile Agency. The platform would be pressurized and manned and would circle the Earth in a polar orbit at 1075 mi altitude. In addition to serving as an observatory, it would provide a base for flights to the Moon.[1]

Shortly after the National Aeronautics and Space Administration was created by Congress in 1958, the space station was awarded priority in the planning program as the next step in space after Project Mercury. In 1959 it was considered as part of an Army space project called Project Horizon that also contemplated outposts on the Moon. As space flight evolved in America, The "next logical step" was deferred time after time by changing priorities that responded to political and economic pressures.

During the 1960s, while the Saturn–Apollo transportation system was being built for the Moon run, NASA's Langley Research Center in Virginia worked unobtrusively on space station design. To the question by Congressional space committee members: what is the space station for? Langley could provide answers. It would enable men and women to learn to live and work in orbit. It would provide a base for microgravity (zero g) research,* a platform for observing the heavens and the Earth, a base for launching deep space missions, and a laboratory for examining the effects of long-term space flight on human beings.

The space station would also enable astronauts to learn the techniques of making rendezvous and docking with another space vehicle in orbit, of extravehicular activity (working outside the spacecraft), and of assembling vehicle systems for deep-space missions.

These purposes shaped the conceptual design for a space station in the 1960s. Engineers at Langley and at the Douglas Aircraft Company sought to encompass them in a project called the Manned Orbital Research Laboratory (MORL). This program became an umbrella for a steadily increasing list of customer demands. As these were presented by prospective users (from government and industry), the design of the MORL space station expanded. There was a demand for facilities to conduct experiments in national defense, Earth sciences, astronomy and future space flight projects.

In its review of MORL evolution, the congressional Office of Technology Assessment said that satisfying user requests would require "hundreds of thousands of man-hours" in orbit. That load implied permanent operations in Earth orbit that would require several stations, or a large station complex. The limiting factor was the cost of logistic support.[2]

The MORL station was defined as a microgravity facility without provision for artificial gravity. However, a centrifuge would be installed in the habitat section for physical therapy if long exposure to microgravity proved to be too debilitating to the crew. MORL inevitably faded away as the next logical step. It lay beyond the financial and technological reach of NASA in a period when national attention was focused on reaching the Moon.

In 1963, as Leroy Gordon Cooper flew 22 orbits in Mercury Atlas 9 to wind up Project Mercury, a more modest space station took shape. The NASA Associate Administrator for Manned Space Flight, Robert C. Seamans, Jr., called for a study of an Earth Orbiting Laboratory (EOL) that could be built within the resources of the space agency while Project Apollo was developing. During this period, the Department of Defense was exploring the potential of a small, manned station as a facility that would enhance national security, presumably by observation. In 1963, the Air Force Manned Orbital Laboratory (MOL) Program was approved as a demonstration to determine how security operations could be carried out by men in an orbiting satellite.

The Air Force trained its own corps of astronauts for 30-day tours in the MOL, a 30-foot cylinder 10

---

* In orbit a space vehicle is falling free around Earth. Consequently, it and its occupants are weightless. But since Earth's gravity holds the vehicle in orbit, it is not entirely absent, but simply not sensed. Thus, the correct term for "zero g" is microgravity.

feet in diameter equipped to support two persons for the month-long mission. At the forward end, the cylinder was attached to a *Gemini B* spacecraft in which the crew rode up to orbit and in which it would return to the launch site, Vandenberg Air Force Base, California, after completing the mission. The laboratory was not equipped for reentry. The MOL program was abandoned in 1969 as a costly duplication of Apollo capabilities, and most of the astronaut trainees were transferred to NASA.

The Office of Space Technology Assessment commented that NASA space studies in the 1960s were characterized by "tension" between the designers of the station and the users.[3] Their concepts did not always mesh. The Office of Manned Space Flight created station designs that did not always accord with the interests of the requirements of the Office of Advanced Research and Technology and the Office of Space Science and Applications. This conflict was to become chronic, for designs were invariably circumscribed by engineering and financial limitations. They frequently fell short of user demands. Late in 1963 the Research and Technology directorate at NASA headquarters asked the field centers to help define "more clearly" the utility of an orbital laboratory for scientific and technological research. This had to be spelled out, not left to an "intuitive judgment" about research potential, an Office of Technology Assessment (OTA) analysis advised Congress. By 1964, the OTA said, the uses of a space station had been defined clearly enough to convince the Office of Manned Space Flight, particularly its Advanced Manned Mission Office, that a space station was timely and necessary. The headquarters unit proposed the creation of an interagency working group that would include the Department of Commerce, the Department of the Interior, and other departments and agencies of government to participate in planning a multipurpose space station.

THE RISE AND DECLINE OF AAP

Long-range planning in the mid-1960s was largely played by ear, under the direction of James E. Webb,

NASA administrator. An energetic, imaginative, and ebullient politician, Webb was President Kennedy's man for the job of running the space agency. What Webb lacked in engineering and scientific qualifications for the post, he more than compensated for with his genius for organization and administration. A lawyer from North Carolina, Webb had been director of the Bureau of the Budget and had served also as an under-secretary of state in the administration of Harry Truman. He knew just about everybody of any consequence on Capitol Hill. He understood the needs of the fledgling aerospace industry, and he sought the support of academia and the scientific community by promoting space science and handing out research grants. If any one executive held the key to the success of Project Apollo it was James E. Webb.

Webb believed in long-range planning, as long as it did not tie his hands. He wanted flexibility in his administrative operations so that he could deal with the conflicting economic and political interests that materialized quickly as America entered the space age. When President Kennedy called for the manned lunar landing, Webb was ready to carry it out. He predicted that its cost might exceed $20 billion. It did. He characterized it as the greatest technical challenge in history. It was. He forecast that the first lunar journey could be made by 1968. By Christmas Day that year, *Apollo 8* was in lunar orbit, with Frank Borman at the helm, reading aloud from the book of Genesis to the people of the blue and white planet that seemed to be hanging in the blackness just outside his window.

Although Project Apollo was funded at the outset as an ad hoc effort to beat the Russians to a manned landing on the Moon, Webb was convinced that it could evolve into a multipurpose transportation system for Solar System reconnaissance and exploration. Beyond the lunar landing program, NASA could amortize the towering cost of Saturn–Apollo by applying the technology to a variety of space programs. Plans to do this were drawn in the Apollo Applications Program (AAP). One application high on the list was a space station, which was still considered the "next logical step," this time beyond the lunar landings.

In 1966 NASA Associate Administrator George

E. Mueller proposed several steps to exploit Saturn–Apollo technology, preliminary to Apollo Applications. The initial one was the Apollo Extension System, which contemplated expanded exploration and research activities on the Moon and in cislunar space, in and beyond near-Earth orbits. Mueller perceived a space station as an outgrowth of Saturn–Apollo technology through the AAP. He contended that a space station from the Saturn–Apollo hardware would amortize some of its enormous development cost. He told the Senate Committee on Aeronautical and Space Sciences, "We can get from Apollo Applications through the development of an orbital space station and then to near planet fly by systems and follow a logical path which goes to planetary exploration."[4]

The purpose of the Apollo Applications Program was to continue developing the Saturn 5–Apollo Transportation System beyond the lunar landing program. Unless the Nixon administration formulated another program, it could be assumed that space activities would follow the Space Task Group plan of 1969. The space station and other elements of the plan could be achieved in Mueller's view through Apollo Applications.

At Houston, Robert R. Gilruth, director of the Manned Spacecraft Center there, argued that a space station derived from Saturn 5–Apollo technology would not fulfill the goal of the "next logical step"—a large, permanently manned space station. Gilruth's view was that the space station should represent a much longer step up in manned space flight development, one comparable to that from Mercury to Apollo. This could not be done through the Apollo Applications design of a station built from the third stage of the Saturn 5. Such a station could house only a small crew and have a limited lifetime in orbit.

The idea of using an empty rocket stage for a space station was suggested in 1960 by Douglas Aircraft Company in a full-scale mock-up of a four-man astronomical observatory that the company built for a display in London. This scheme was realized a decade later as the Apollo Applications workshop inside the liquid hydrogen tank of the S4B, the third stage of the Saturn 5. It evolved as Skylab. Although it was not the modular, multiunit

structure Gilruth advocated, it was within the state of the art and the state of NASA's appropriations.

As the lunar landing program got under way, the likelihood that the administration would commit the nation to a space station faded. The public, fascinated by the early landings, exhibited little interest in a station, an attitude that was reflected in Congress. The prospect of using the Saturn 5 to explore the planets was put down by critics as unrealistic in view of the escalating costs of the war in Viet Nam and the fiscal demands of federal entitlement programs. The price tag for Apollo alone had risen to $3 billion a year, more than half of NASA's budget.[5]

Originally programmed to make 10 lunar landings, Project Apollo was cut to seven because of escalating flight costs. Unexpectedly, public interest, which had bordered on euphoria during the first two missions, diminished as the sense of novelty wore off. The reality of the lunar adventure was less interesting than the challenge had been. Despite the exposure to danger on a new and unknown world, astronaut activity on a monotonous landscape lost its fascination. It was hard work. The essential discovery that the Moon was the Rosetta Stone of terrestrial planet evolution was not revealed until the missions were completed and the results were presented at lunar science conferences in Houston.

At the first lunar science conference in February 1970, George M. Low, NASA Deputy Administrator for Manned Space Flight, announced that the tenth mission, Apollo 20, would be cut. The Saturn 5–Apollo reserved for it would be used to launch a space station if that project was approved. In September 1970, the Apollo 18 and 19 missions were canceled. Of the seven remaining missions, the landing was aborted on only one, Apollo 13, after the oxygen storage tank blew up en route to the Moon. The crew returned to Earth on a circumlunar trajectory, using oxygen supplies in the lunar module to sustain them on the run home. The other six missions were successful in landing astronauts on the lunar surface, the last of them, Apollo 17, in December 1972.

Aside from its remarkable success as an interworld transport, the Saturn 5–Apollo system proved to be too expensive, in terms of NASA funding in the 1970s, for routine, repetitive flights. At the Langley Research Center, a team of engineers created designs for a reusable space shuttle, a vehicle that could be flown again and again instead of being thrown away after every flight; but one that because it had wings would be forever restricted to flights between the surface of the Earth and low Earth orbit. In time, the space shuttle would supplant Saturn–Apollo. NASA would retreat from the Moon to low Earth orbit for the balance of the century with a commitment to redevelop the technology of manned space flight to achieve airline reusability, after a decade of launching men by expendable rockets in ballistic spacecraft.

THE LOST HORIZON

NASA's colorful leader, Administrator James Webb, resigned in the closing weeks of 1968 as *Apollo 8* was prepared to make the first flight to lunar orbit. In December *Apollo 8* achieved its lunar orbit mission and an American victory in the space race was in the bag. Webb was succeeded by Thomas O. Paine, a research engineer and former executive of the General Electric Company. Paine had joined NASA in 1968 as deputy administrator.

Amid confident preparation for the first lunar landing, Paine requested $60 million to start the space station program, but the Bureau of the Budget rejected it. As acting administrator, Paine pushed Webb's development policy but was unable to get station funding. Although a Democrat, Paine was nominated in March 1969 by President Nixon to be NASA administrator and was confirmed by the Senate. He remained at the head of NASA for only 20 months, resigning July 29, 1970. As a Democrat holding a high post in a patronage-hungry Republican administration, he feared that he might become a political liability for the space agency's funding efforts. "I was in the position of a general managing a retreat," he said.[6]

Paine was succeeded by a Republican, James C. Fletcher, the president of the University of Utah and a former aerospace industry executive. During his brief term in office, Paine had been rebuffed in his efforts to win a commitment for a space station. As early as February 1969, he had appealed directly to Nixon to support the station before the momentum of the Apollo program was lost. Paine's appeal preceded establishment of the Space Task Group, which was organized by the Nixon administration on February 13, 1969, to design the post-Apollo program. In its review of space station development, the Office of Technology Assessment reported that some observers believed that Paine wanted a space station commitment from the president before the post-Apollo plan was promulgated.

Although the Space Task Group had debated the space station as the next goal in space, the view was not unanimous. The space science and technology subcommittee of the President's Science Advisory Committee advised that the station was not an urgent necessity. "From a programmatic standpoint, the arguments in favor of early action appear very weak," the advisory committee stated.[7]

The staff of the Space Task Group rejected the station as an early commitment, and most of the committee members did not support funding for more rapid progress toward launching the station. However, the committee stated that this position did not imply that the majority opposed the station as a new goal, but those taking that position did not want to prejudge the outcome of the Space Task Group's deliberations.

In the Cabinet, the State Department offered support for the station, citing its value in promoting international cooperation. The Department of Defense had no interest in the station per se but indicated a concern with the need to develop lower-cost transportation than the Saturn–Apollo system provided. In space station studies in 1970, it was determined that Saturn–Apollo system would provide logistics for the station. The Saturn 5 would boost the station into orbit, and the smaller Saturn 1B launcher would send the crew aloft to man it in the Apollo spacecraft.

There were interests in the space industrial community that regarded the development of lower-cost space transportation as a higher national priority than the space station, as did the Department of Defense. By mid-1970, said the Office of Technology Assessment, "it was clear that in the eyes of the space subgovernment outside NASA, the shuttle program was a more attractive investment than the space station. . . . by the end of the year, the space station had been eroded to the status of a conceptual study."[8]

A reversal of NASA policy had occurred. In the summer of 1969, NASA had let two contracts for the definition of space station design (Phase B in project development, a step up from the Phase A conceptual study). At congressional hearings during 1970, NASA had presented the space station as its main post-Apollo objective. Suddenly (from the viewpoint of observers), Phase B work at the Manned Spacecraft Center in Houston and the Marshall Space Flight Center in Huntsville was terminated July 29, 1970 by Charles Mathews, Manned Space Flight Deputy Associate Administrator.

Phase B studies had begun in September 1969. The earlier Phase A studies of 1967–1968 had depicted a station 33 feet in diameter (to fit on the Saturn 5) for a crew of six to nine astronauts, with a two-year lifetime in orbit. Launched by Saturn 5, it would be serviced by Apollo, boosted by Saturn 1B. As mentioned earlier, this projection was regarded as too conservative. Gilruth at Houston and Wernher von Braun at Huntsville urged a more ambitious station, a much bigger one, that would arouse public interest and awe, as the lunar landings had done. Gilruth argued that several Saturn 5 launches could put a million pounds in orbit, providing the hardware for a large orbital base.

In Phase B, responding to these urgings, the space agency looked toward the development of a 100-man space base, starting with an initial 12-man station by 1975. The project work statement described the station as a general-purpose laboratory for scientific and technological experiments. It would eventually be developed to provide a launch capability for extended space exploration. The initial station, as defined by Phase A, would remain 33 feet in diameter. It would be launched into a 270-nautical-mile orbit (310.5 statute miles) at an inclination of 55° to the equator.

The initial 12-man station would operate in microgravity (zero G), but an assessment of the need for artificial gravity would be made. This could be done by tying a counterweight on a long tether to

the station and spinning them to provide the pseudogravitational effect. The 33-ft-diameter station would not only would provide the core of an extended space base but also could be configured as the habitat module of an interplanetary transport launched from low Earth orbit to Mars.[9]

A continuing problem during Phase B was the integration of station design with experiment facilities. By June 1970, this problem and others had caused the station to slip to 1977 completion. The cost of the program was estimated at $8 to $15 billion, including development and 10 years of orbital operations. Although the station would be put into orbit by Saturn 5, it could be serviced by the shuttle instead of by the Saturn 1B–Apollo system. The cost of the shuttle vehicle was not included in these estimates.[10]

By the summer of 1970, following Paine's resignation, NASA leadership "grudgingly" made the shuttle its top-priority program, the Office of Technology Assessment reported.[11] Department of Defense support for the shuttle provided leverage for this priority shift; the department had expressed disinterest in the space station as long as NASA invited international participation in it. Paine's resignation and the termination of Phase B space station studies by Charles Mathews, both of which occurred July 29, 1970, underscored the revolution in manned space flight programming that had occurred at NASA headquarters. The space station was out. The space shuttle was in, provided its development cost could be shaved low enough to satisfy the budget.

The significance of this fundamental policy switch was that the Saturn 5–Apollo transportation system also was terminal, although it would not become extinct for several years. This in-house revolution was costly. The United States lost its powerful and reliable heavy launch vehicle. The time would come when the Saturn 5 would be desperately needed to match Soviet competition in the development of space bases in Earth orbit.

The transition from Saturn 5–Apollo to the shuttle required a fundamental revision in space station design. The 33-foot-diameter core station designed for the Saturn 5 was rendered obsolete. It was too big to be lifted into orbit by the shuttle, with its 15-

foot-diameter cargo bay. From 1970 on, the future U.S. space station would be constructed with 14-foot-diameter cylinders up to 45 ft long providing pressurized living and working space. *Skylab,* the last hurrah of Saturn 5 technology, was already in development as the switch in space transportation systems was approved by the Nixon administration. Work on *Skylab* continued, however, at the Marshall Space Flight Center, even though the Apollo Applications concept the station represented was officially a dead end.

Space station studies were refocused in 1972 on a small research and applications module known by its acronym, RAM. It was sized so that it could be carried into orbit by the shuttle. RAM, also known as Sortie Lab, evolved into *Spacelab,* a workshop consisting of one or more pressurized modules carried in the cargo bay of the shuttle orbiter with open pallets. The pressurized workshop was linked to the crew cabin by a tunnel and became an organic part of the orbiter in flight. The workshop volume was about one fourth that of *Skylab,* and its flight time was 10 days—compared with 28, 59, and 84 days in *Skylab. Spacelab* was built by the European Space Agency as a joint venture with NASA for scientific and technological research in orbit.

A MARTIAN MANIFESTO

Although the Nixon administration turned down Paine's space station initiative, NASA persisted in recommending stations in low Earth and geosynchronous orbits in its far-ranging report "America's Next Decades in Space." The report reflected Paine's outlook and expressed a boldness that impressed Vice-President Spiro Agnew. Agnew was impressed by the realization that a landing on the Moon would be a hard act to follow. This was not a time to pull back. He told a news conference at the launch of *Apollo 11,* July 16, 1969, that, "we should articulate a simple ambitious, optimistic goal of a manned flight to Mars by the end of century." Paine, in his first year as NASA administrator, agreed. He proposed that the Space Task Group adopt an early Mars mission as the main focus of its planning

recommendations. As NASA administrator, he encouraged agency planners who were drafting recommendations to include a manned Mars mission during the 1980s into NASA's long-term planning proposal.

In 1969, the perception of Mars was not one of a garden spot. Prior to the Space Task Group recommendations, only indistinct images of Mars close up had been seen by humanity. They were amorphous photographs of the surface, predominantly the south polar region, from *Mariner 4,* which flew by the planet July 14, 1965. Radioed to Earth, the *Mariner 4* pictures portrayed a dismal vision of a heavily cratered, lunarlike surface, rimmed with frost. Other data indicated a carbon dioxide atmosphere so thin that explorers would have to wear space suits to survive on the surface. Two more Mariner spacecraft were en route to Mars in 1969, *Mariner 6,* due to fly by at closet approach July 31, and *Mariner 7,* due August 6. Their results would not improve the lunar-like scenes of *Mariner 4* substantially. Later reconnaissance would change this picture of Mars as another Moon, but the Mariner images hardly suggested the promised land.

The final Space Task Group report recommended that the United States "accept the long range option or goal of manned planetary exploration with a manned Mars mission before the end of the century as the first target." The report was submitted to President Nixon in mid-September 1969.

A Mars mission appeared to be a bold undertaking at a time when Apollo astronauts had just started to collect rocks on the Moon, but it had little impact on the White House. The Office of Technology Assessment noted that six months passed before the president responded to the Space Task Group report. The response, the Office noted, was noncommittal.

SKYLAB

Prior to 1969, post-Apollo manned space flight was projected on the basis of the Saturn–Apollo Transportation System, which had evolved in the 1960s through Mercury, Gemini, and Apollo and the invention of the Saturn launchers. A similar evolution of ballistic manned spacecraft was going on in the Soviet Union with *Vostok*, *Voskhod*, and *Soyuz* spacecraft. The United States won the race to land men on the Moon with the *Saturn 5*. The USSR lacked a launch vehicle of comparable power in that period, but it successfully carried out unmanned lunar exploration with automated rovers and sample return vehicles.

In the mid-1960s, NASA post-Apollo planning was based on a decision "to use, modify and expand present Apollo system capabilities rather than move toward whole new developments."* This was the rationale of the Apollo Applications Program and its only product, the *Skylab* space station.

Skylab was developed from a rocket stage rather than from a manned satellite as the Soviet *Salyut* space station was in the same period. The basic idea, as mentioned before, had been proposed by the Douglas Aircraft Company. During Saturn rocket construction at the Marshall Space Flight Center, Wernher von Braun, center director, ordered a study of using the liquid hydrogen tank of the S4B, the third stage of the Saturn 5, as an orbital workshop. This project was approved by George E. Mueller, NASA's chief of manned flight. The S4B was 58 feet, 7 inches long and 21 feet, 8 inches in diameter. The liquid hydrogen tank, its main component, was 48 ft long. Without its load of 63,000 gallons of liquid hydrogen, the tank provided a volume of 10,000 cubic feet. The S4B also contained the liquid oxygen tank holding 20,000 gallons. It could be used empty for storage or waste disposal.

By 1965 the utility of converting the S4B to a workshop had been established at Marshall. Initially, it would be launched as the second stage on the smaller Saturn 1B and drained of liquid hydrogen fuel and vented in orbit. An Apollo crew would then enter the empty tank and convert it into a workshop with living quarters.† Previously installed structural components, including metal flooring, would not be affected by the hydrogen

fuel and could simply be emplaced or assembled. Food, water, air, and other consumables and supplies and equipment would then be loaded in the tank. This phase of development was called the "wet workshop."

With the cutback in lunar landing missions, a big Saturn 5 launch vehicle became available in lieu of the Saturn 1B. The S4B was the third stage of the Saturn 5. The first two stages could easily lift it into orbit, dry and fully equipped and provisioned. Skylab then was to be realized with a "dry workshop." Mueller had proposed a plan to convert an Apollo spacecraft into a manned solar observatory. It was developed as the Apollo Telescope Mount and was attached to the Workshop.

Several Skylab stations were planned in low Earth and geostationary orbits, but only one was built after headquarters decided to abandon the Saturn 5. *Skylab* was a massive cluster. With Apollo docked to it, the space station, with its 48-ft workshop, 22 ft in diameter, was 117 ft long and had a mass of 199,750 lb. It contained a habitable volume of 12,700 cubic feet, about four times that of the Soviet Salyut 6 space station. There were 9550 cubic feet of pressurized volume for crew quarters, laboratory, and storage in the workshop. The big tank supported the main solar power wings and contained cold gas tanks and thrusters for secondary attitude control.

Attached to the workshop in tandem was the Air Lock Module and Docking Adapter. The Air Lock contained the communications equipment, data-processing machines, environmental control equipment, and electric power systems. The module also provided means to go in and out of the pressurized station volume for repairs outside. The Multiple Docking Adapter provided Apollo docking ports. It also housed the control console for the Apollo Telescope Mount and controls and sensors for viewing Earth. Atop the Apollo Telescope Mount was a windmill array of four photovoltaic panels that provided electric power to supplement the power supplied by the main power wings on the workshop. At least one advanced model of *Skylab* had been planned, but it was abandoned with termination of Saturn 5 production and Apollo Applications. *Skylab* was not fully utilized. It was occupied for only

---

*NASA announced this decision as the rationale for the Apollo Applications Program on January 26, 1967, at a briefing in Washington. The significance of the announcement was muted by the *Apollo 1* fire, in which the test crew perished the next day at Cape Canaveral.

†The crew was limited to three, the crew complement of the Apollo command module.

three missions by three-man Apollo crews for periods of 28, 59, and 84 days—a total of 171 days, 13 hours, and 14 minutes, between May 14, 1973, and February 8, 1974. A substantial part of crew flight time was taken up by the necessity of making repairs, inside and outside the station. The workshop was damaged at launch May 14, 1973, when one of the solar power wings was torn off by the premature deployment of a meteorite shield that also protected the workshop from heating by the sun. By heroic effort, the crews of the first and second missions were able to restore part of the workshop power system and rig up a sun shade, as well as make other repairs.

After the third crew departed February 8, 1974, *Skylab,* still usable, was abandoned. Without a revisit by an Apollo spacecraft, the station lacked propulsion to correct a decaying orbit and became a derelict. It reentered the atmosphere and crashed in the Indian Ocean off the southwest coast of Australia near Perth July 11, 1979.

MOSC

With the termination of Saturn 5–Apollo, the United States became dependent on a single system, the troubled shuttle program, for manned access to

Figure 1.1. Damaged during launch May 14, 1973 but usable, here is Skylab, America's first space station as it looked to the crew of an approaching Apollo spacecraft. It was occupied by rotating crew of three astronauts for three missions of 28, 59, and 84 days. Abandoned after the third crew departed February 8, 1974, the big station reentered the atmosphere July 11, 1979 and crashed into the Pacific Ocean off the west coast of Australia. This photo shows how the sun shield was rigged (the white cover) by astronauts to replace an aluminum canopy that was ripped off with one solar cell wing during launch. (NASA)

orbit and the construction of a space station. The transition in space transportation systems to the shuttle resulted in a six-year hiatus in American manned space flight (1975–1981). The Soviet Union was thus able to surpass the United States in maintaining a manned presence in orbit by ongoing space station development and in studying the physiological effects of long-term crew exposure to microgravity. These effects on bone and muscle tissue and on the cardiovascular system had been observed on *Skylab,* but it was essential to understand the effects over much longer terms of exposure before manned interplanetary flight could be seriously programmed.

In terms of manned flight experience, the trade off in transportation systems put the United States behind the Soviet Union. This was the result of more than eight years down time in manned space flight, including the grounding of the shuttle fleet for the redesign of solid rocket booster seals for the shuttle after the *Challenger* accident of January 28, 1986.

Financial benefits of the tradeoff proved chimerical after the shuttle began flying in the 1980s. Operating costs were higher than predicted, allowing the European Space Agency to compete successfully against the shuttle for satellite payloads with the expendable Ariane launcher. After 24 missions, the shuttle program failed to prove its cost effectiveness, as compared with expendable launchers, or its utility as a research facility and observatory. *Skylab* had successfully demonstrated the scientific and technological value of an orbital research station. The Apollo Telescope Mount, its solar observatory, produced thousands of high-quality photos of the sun that solar physicists have studied for years.

For the first time NASA acquired physiological data on the effects of exposure to microgravity on bone and muscle tissue for periods of 28, 59, and 84 days. The results showed that measures, principally exercise, could be taken in flight to reduce neuromuscular deterioration and bone loss and speed up full recovery of crews on their return to Earth. Whether artificial gravity would be necessary to maintain crew health on voyages of eight months or a year to Mars could not be determined

from the *Skylab* experience, but it was clear that regular exercise would be an essential feature of the crew routine aboard orbiting space stations.

With the end of *Skylab,* NASA planners began to search for a low-cost alternative. A new study was inaugurated in 1975 called the Manned Orbital Systems Concept (MOSC). The engineering details were worked out by McDonnell Douglas Astronautics under the technical supervision of Marshall Space Flight Center. The MOSC study analyzed requirements for a cost-effective (a dominant requirement) orbital facility capable of supporting manned operations longer than the seven- to ten-day flights projected for the shuttle with *Spacelab* in the cargo bay. The new facility was scheduled to commence operating by the end of 1984.

Instead of a general program of scientific and technological research, the main purpose of MOSC was the study of Earth resources. It was a space station rationale that responded for the first time to concerns about climate changes, overpopulation, and limits to growth. MOSC turned 180 degrees from the idea of dealing with the limits to growth by exploiting extraterrestrial resources of matter and energy, a popular rationale for space programs in the early 1970s. MOSC shifted the focus of space station development to conservation of planetary resources and monitoring the terrestrial biosphere. This shift was consistent with the retreat to low Earth orbit and its parochial programmatic implications.

The MOSC study produced a four-man, modularized station, based on *Spacelab* design with pressurized modules and pallets. McDonnell Douglas estimated the cost of developing the small facility at $1.2 billion. The MOSC effort did not continue beyond the Phase A conceptual stage. In the fall of 1975 NASA shifted back to a more ambitious program that emphasized the construction of facilities in orbit and the assembly of a large, modular base. The new direction the study was taking called for scientific and technological research facilities, the development of solar energy collectors, solar mirrors as power sources, the capability of retrieving satellites for repair and for docking vehicles returning from lunar or interplanetary missions. However, emphasis of the enlarged study remained on Earth observation and monitoring.

During the mid-1970s, while the shuttle was in development, Fletcher set up a study group to identify major objectives in space for the remainder of the twentieth century. The task was assigned to headquarters and center personnel. It resulted in a revival of 1969 NASA plans entitled "Outlook for Space," published in 1976. Although a principal thrust of the work was development of Earth-oriented scientific investigation, the orientation the Nixon administration preferred, in a calm and undramatic way it projected a rationale for the exploration of extraterrestrial as well as terrestrial resources. It balanced the need for the collection of data that would improve weather and climate forecasting, crop production, delineation of water resources, and early warning of environmental problems with proposals for developing solar power satellites and investigating a means of disposing of nuclear wastes in space.

Beyond these measures, the "Outlook" recommended a permanent space station "to fully exploit the zero gravity environment of space for basic research." It would help develop full understanding of man's ability to live and work in space for extended periods, the work added. "We need to know in particular what are the physiological and medical consequences of long duration weightlessness, whether any protective measures such as artificial gravity may be required and what are the safe human tolerance limits to ionizing radiation from the sun, the galaxy and space nuclear power systems." This statement expressed one of the most direct and comprehensive concerns for human well-being in space travel in NASA literature. It was concise and definite, and up to this writing, the questions it poses have not been resolved by the space agency.

Beyond Earth orbit, the "Outlook" said, a manned base on the Moon might well be considered for developing mineral resources. Lunar aluminum, for example, could be processed for the construction of solar power stations, which could be launched from the Moon into geostationary orbit with one twentieth of the energy cost of launching them from Earth. "Thus a lunar industry capable of

manufacturing space power stations could be worth tens of billions of dollars a year," the "Outlook" estimated. Although the "Outlook" did not advocate a space station strongly, it concluded that, "With the shuttle system giving us comparatively low cost access to space on the one hand and the economies which could be realized from the use of a permanent space facility on the other hand, the construction of a permanent space station appears to be the next logical step for the manned flight program. . . ."

At the end of the 1970s, engineers at the Marshall and Johnson space centers began their own space station studies in the absence of any specific direction for headquarters. Its hands full with the shuttle, the Carter administration produced no space initiatives. Fletcher had been succeeded as NASA administrator in 1977 by Robert A. Frosch, a physicist and former assistant secretary of the Navy for research and development and associate director of the Woods Hole Oceanographic Institution. Frosch left NASA in 1981 to become president of the American Association of Engineering Societies as the administration changed.

Workers at NASA's Kennedy Space Center in Florida characterized Carter's space policy as "no go." There were no new starts. Carter's attitude toward space activity was expressed by a White House fact sheet on the U.S. Civil Space Policy: "It is neither feasible nor necessary at this time to commit the United States to a higher challenge space engineering initiative comparable to Apollo." Presumably the author of this policy did not recognize development of the space shuttle as a high-challenge initiative.

The administration's failure to institute other space activities aroused criticism in Congress. During hearings on the NASA budget for the 1980 fiscal year, Senator Adlai E. Stevenson (D–Ill.), chairman of the Senate Subcommittee on Science, Technology, and Space, told Frosch, "I sense that the first priority of this administration is to study, study and re-study. That's why this administration is such a small achiever."[12]

NASA centers responded to shuttle preoccupation and the lack of new initiatives at NASA headquarters with new initiatives. Engineers at the

Marshall Space Flight Center began a study in 1978 for an unmanned science and applications space platform that would evolve into a manned system. At the same time, engineers at the Johnson Space Center started planning a major new space station.

Increasing Soviet manned presence in orbit, demonstrated by the Soviet *Salyut* space station system, was generating concern among NASA engineers and scientists. In 1979, the Johnson Space Center engineers made a study of a facility in geostationary orbit, 22,300 miles above the equator, called a Space Operations Center (SOC). They believed that space station services for satellites and instrument platforms would one day be required there. This study produced a new scenario: an orbital station constructed from several modules that would be launched by the shuttle. The station would accommodate a crew of four to eight persons and

would have facilities to support flight vehicles and construction activities. The cost was estimated at $2.7 billion over a decade.

These center projects appear to have been inspired by the lack of planning direction at headquarters. Subsequently, grassroots lunar and planetary exploration initiatives appeared at the centers, neither pushed nor discouraged by official policy. The space station projects had no official support in the Carter years. The Office of Technology Assessment concluded in its 1980 report that there was no necessity for developing a multipurpose space station infrastructure soon. Such was the consensus among federal advisory agencies.

The influential Space Science Board of the National Research Council/National Academy of Sciences asserted that nearly all the space science projects forecast for the next 20 years could be carried out without a space station. All the missions

Figure 1.2. The Space Shuttle Columbia, shown as it landed December 8, 1983 at Edwards Air Force Base, California after a 10-day mission, carried Spacelab in its cargo bay. In the next decade it will be flown as a construction base for the Freedom Space Station. Launched April 12, 1981, it is the world's oldest manned spacecraft in continuous flight service. (NASA)

these projects required could be done by the shuttle and spacelab, by satellites, and probes. In its report of September 15, 1983, the board said that, although such a station might eventually be useful, there was no need for it in the next two decades. However, a space station could serve "as a very useful facility in support of future space science activities," conceded Thomas M. Donahue, the board chairman, in a letter to James M. Beggs, who became NASA administrator in 1981.[13] Another advisory group of the National Research Council, the Space Applications Board, expressed interest in the use of an orbital facility to service free-flying platforms, launch satellites from low Earth orbit to geostationary orbit, repair satellites, and provide equipment for materials processing in orbit. NASA's Solar System Exploration Committee, a group of

space scientists, fully endorsed the space station. It would be a base for the assembly and launch of deep space probes and a place to store a returned soil sample from Mars until its contamination hazard could be determined.

It was Beggs who rekindled administration interest in the space station as a national necessity for matching the Russian manned presence in Earth orbit. He lobbied hard for the station against opposition from influential members of the science establishment, including George Keyworth, the president's science adviser. On May 20, 1982, Beggs formed a Space Station Task Force to draw an initial space station plan. NASA then contracted with eight aerospace firms to identify user requirements in space science and applications, technology development, commercial activities, and na-

Figure 1.3. This artist's rendering depicts a space platform concept as visualized by Marshall Space Flight Center engineers in 1981. It was designed as a man tended structure with a solar wing power system. Deployed in orbit by the shuttle, it would provide some of the experiment facilities of a space station. (NASA)

tional security. Canada, the European Space Agency, and Japan were interested in possible participation in the program and made corollary studies.

At NASA field centers, working groups were organized in the summer of 1982. Studies were begun on platforms and two supporting vehicles, an Orbital Transfer Vehicle that would fly between low Earth, geostationary, and lunar orbits and an Orbital Maneuvering Vehicle that would move payloads in the vicinity of the station.

On July 4, 1982, the Space Shuttle *Columbia* completed its fourth test flight. When it landed at Edwards Air Force Base, California, it was pronounced operational. In the view of Administrator Beggs, it would serve as the construction vehicle for assembling the space station in low Earth orbit.

The space station at last was moving toward realization, but as was clear to its supporters, it required an authorization from the President of the United States. In 1983, President Reagan moved toward such a pronouncement. He directed the Senior Interagency Group for Space, consisting of cabinet-level departments and agencies, to review NASA's space station concept. On November 15, 1983, the director of the Space Station Task Force, John D. Hodge, appeared before the Senate Subcommittee on Science, Technology, and Space to explain the new plan: "with the shuttle having transitioned into initial operational status," he said, "it is now time to plan for the next logical step in the exploration and exploitation of space. . . . There have been significant space station studies since the

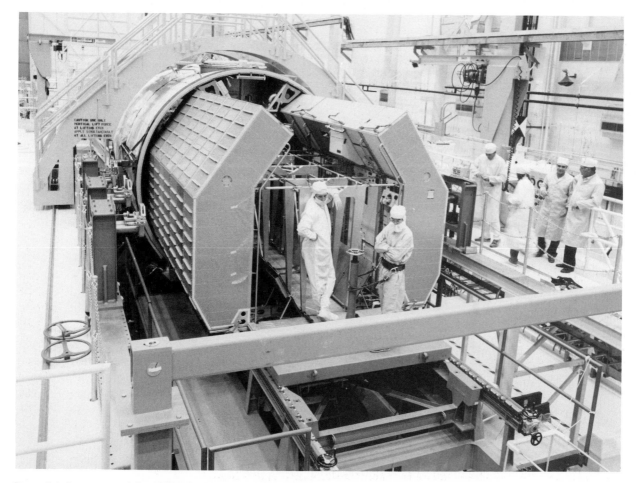

Figure 1.4. During part of the 1980s, the European Space Agency's Spacelab provided laboratory facilities for experiments aboard the shuttle. In this photo, experiment racks and floor are being inserted in the shell of the pressurized module that flew in the cargo bay of the shuttle orbiter Challenger in 1985.

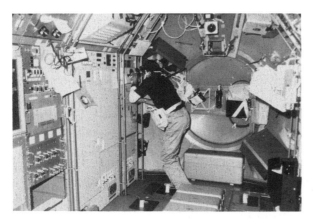

Figure 1.5. Lodewijk van den Berg of EG&G Energy Management, Inc. observes vapor crystal growth in a French experiment aboard Spacelab 3 in the Challenger cargo bay on shuttle mission 51B April 29–May 6, 1985.

Skylab missions 10 years ago. In addition to the science and application laboratory functions performed on board Skylab, today's concept includes the servicing of free flying satellites and platforms, the assembly of large structures in space, the basing of upper stage vehicles and propellants and promising commercial manufacturing and materials processing missions."

These purposes would shape the architecture of the new space station. The station would be the nation's third major manned space enterprise, following the lunar landings 1969–1972 and the invention of the reusable space shuttle, 1972–1981.

A new design emerged on the drawing boards of NASA architects shaped by the shuttle's 60×15-foot cargo bay and by user requirements that had accumulated for more than a decade. The projected station would carry instruments to acquire data on Earth's environment, on fields and forces radiating from the sun and other stars, and on human physiological responses to microgravity; it would provide facilities for materials processing in microgravity. The station would also be equipped to serve as a repair base for space vehicles. Satellites and probes would be assembled and tested there before being deployed in Earth or lunar orbits or on interplanetary voyages. The station could be a staging depot for large, interplanetary vehicles.

Late in 1983 Administrator Beggs presented the case for a permanently manned space station at a cabinet meeting. The Inter-Agency group had failed to reach a consensus on the station and its members offered conflicting views. At the beginning of 1984, the way was open for the space station as the only new major project on the national space agenda. The shuttle was by then deemed by NASA to be a fully operational system. It was being flown principally as a satellite carrier to low Earth orbit, replacing *Delta*, *Atlas*, and *Titan* expendable boosters. Its main function, to develop and service a space station, as conceived by the Space Task Group, remained unfulfilled. Still, it was part of a system to expand the human presence beyond Earth and into the Solar System. The space station was the other part. Without both parts, humans were confined to low Earth orbit. They could not even reach the Moon, 15 years after Apollo had done so. By

1984, the space station no longer competed for funds with Apollo or the shuttle, although some investigators feared that it would swamp other initiatives. By 1984 the space station was NASA's only game in town.

President Reagan acted. In his State of the Union message of January 25, 1984, he directed NASA to develop a permanently manned space station within a decade. By the summer of 1984, NASA headquarters had assigned preliminary station design studies to four field centers. The Marshall Space Flight Center at Huntsville, Alabama, was assigned to define the pressurized modules with a common design. These were the crew habitat, laboratory, and logistics modules, cylindrical in shape, 14 feet in diameter and 40 to 44 feet long. Marshall would manage the environmental control and the propulsion systems.

At Houston, the Johnson Space Center was ordered to define the structural framework, a truss, to which the modules would be attached (after being carried up to orbit separately in the shuttle cargo bay). It would also integrate mechanical and power systems into the structure, which would be put together by astronauts like a huge erector set, and would design the interfaces between the station and the shuttle and the remote manipulator system. Johnson would also manage the attitude control, thermal control, communications, and data management systems and would equip the ward room and galley in the crew module.

The job of designing the electrical power system was given to the Lewis Research Center at Cleveland, Ohio. The primary power system was considered to be solar, but alternative nuclear and chemical (fuel cell) systems would be evaluated. Later, a steam turbine generator, powered by solar heat, was proposed.

Goddard Space Flight Center at Greenbelt, Maryland, was to define the laboratory module and design automated, free-flying experiment platforms. One platform would be launched into polar orbit; the other would co-orbit with the station, both transmitting data from their experiments to the tracking and data relay satellite (TDRS) system that NASA planned to have established in orbit by the end of the 1980s. In Florida, the Kennedy Space

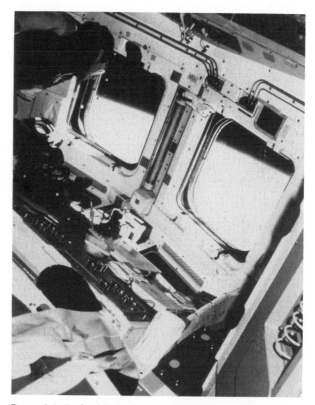

Figure 1.6. Payload Specialist van den Berg peers out Challenger's aft flight deck window during the 51B flight. Earth's horizon and air glow are faintly visible in this photo. Earth viewing will be carried out extensively in the Freedom Space Station in environment studies. (NASA)

Center would perform shuttle cargo bay loading and launch operations for the station.

During 1984 NASA held discussions with Canada, the European Space Agency, and Japan about their participation in the space station. With the prospect that the station would be internationalized, the Department of Defense deferred its interest in it. In the meantime, preliminary responses from the contractors indicated that the station and its support vehicles would be the largest engineering project of the space age, outdoing Project Apollo in complexity, cost, and potential. It had to be if NASA was to pursue the dream of establishing a human presence beyond Earth.

SPACELAB

Pending a presidential decision on a space station, NASA made an agreement with the European Space Agency (ESA) for the development of short-term laboratory modules, collectively called *Spacelab*. A pressurized, cylindrical module 23 feet long (including end cones) and 13.31 feet in diameter was the principal laboratory structure. It consisted of two segments shaped like a bass drum that could be flown separately or joined in the "long module" configuration.

The pressurized modules were assembled in Bremen, West Germany, by VFW–Fokker/ERNO. The long module, about the size of an old-fashioned streetcar, would be flown separately or with extensions in the form of unpressurized, U-shaped pallets on which sensing instruments and experiments could be mounted. The interior of the pressurized module was fully equipped as a laboratory and experiment station.

Spacelab was designed to fit neatly in the cargo bay of the space shuttle orbiter. The first *Spacelab* mission was flown by Columbia, for 10 days, 7 hours, with a crew of six following launch on November 28, 1983. It carried 72 experiments. Three subsequent *Spacelab* missions were launched by *Challenger* April 29, July 29, and October 30, 1985, all for seven days. The range of investigations included high-energy physics, microgravity, astronomy, solar physics, and life sciences.

Figure 1.7. A sled on rails was designed by European Space Agency engineers for the investigation of motion sickness and accelerations that produce it in microgravity. Built for ESA's Spacelab, the sled could be used also for microgravity adaptation research in a space station. (ESA)

Figure 1.8. The Space Shuttle Columbia with the European Space Agency's Spacelab aboard is depicted in this cutaway painting. (ESA)

BIRTH OF A STATION

AFTER a quarter century of planning, America's second attempt to develop a permanent manned space station began to move forward in 1984 as a national commitment. As *Skylab* had evolved from the Saturn–Apollo Transportation System in the 1960s, the station of the 1980s was entirely the product of the Space Shuttle Transportation System. Its basic design of an array of cylindrical modules attached to a stabilizing truss was determined by the dimensions of the shuttle payload bay that would carry the modules into orbit. The single-tank architecture of *Skylab,* the earlier *Manned Orbital Laboratory, Spacelab,* and the Soviet *Salyut* stations was replaced by a complex of four large cylinders equipped as a habitat and laboratories centered on a long structure that looked like a bridge—a bridge in space.

The bridge was a truss 262.5 ft long. It formed the trunk of a tree from which modules, fixtures, and spacecraft would become branches to be interconnected by nodes and tunnels through which flowed air and power. At each end of the great boom, glistening arrays of photovoltaic cells would grow to become huge, rectangular, leaflike struc-

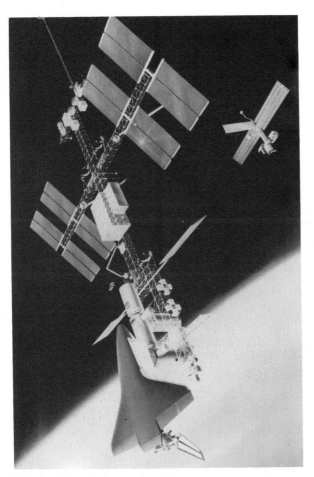

Figure 2.1. Following President Reagan's directive in January 1984 to develop a space station, the Johnson Space Center developed this "reference configuration" of the station by September 1984. A shuttle orbiter is shown docked with the lower part of the structure and photovoltaic solar panels are extended at the top. Off to one side, a free flying platform is shown. (NASA)

tures, transforming sunshine into electrical energy for the tree. The tree was to be planted before the end of the century. In time, it would grow to provide a staging base for flights to the Moon and planets.

Nine months after the station was authorized by the president, NASA called for preliminary design and definition proposals (Phase B) from industry. Proposals were solicited for four separate work packages. By mid-November 1984, NASA headquarters announced receipt of 13 proposals from teams of aerospace contractors.

Contracts for the preliminary designs were awarded April 19, 1985. Work Package 1 called for the design of pressurized common modules equipped for use as a habitat, as laboratories and logistics transport, as environmental control and propulsion systems, as accommodations for Orbital Maneuvering Vehicles and Orbital Transport Vehicles, and for the outfitting of the modules. This work package was to be managed by the Marshall Space Flight Center. Contracts for the Phase B definitions and designs were awarded on a competitive basis to the Boeing Aerospace Company, Seattle, Washington, and to Martin Marietta Aerospace, Denver, Colorado.

Work Package Two, to be managed by the Johnson Space Center, called for definition and preliminary design of the structural framework on which the modules and other elements of the station would be attached; the interface between the station and the shuttle; the Remote Manipulator System (RMS); attitude control, thermal control, communications and data management systems; plans for equipping a module with sleeping quarters, wardroom and galley and plans for extravehicular activity (EVA). Competitive contracts were awarded to McDonnell Douglas Astronautics Company, Huntington Beach, California, and to Rockwell International, Space Station Systems Division, Downey, California.

Work Package Three, to be managed by the Goddard Space Flight Center, Greenbelt, Maryland, called for preliminary definition and design of the automated, free-flying platforms; provisions for instruments and payloads to be attached externally to the station; and plans for equipping a module as a laboratory. Competing contracts were awarded to RCA Astro Electronics, Princeton, New Jersey, and to the General Electric Company, Space Systems Division, Philadelphia, Pennsylvania.

Work Package Four, to be managed by the Lewis Research Center, Cleveland, Ohio, covered the design and definition of the electrical power generating, conditioning, and storage systems. Contracts were awarded to the Rocketdyne Division of Rockwell International, Canoga Park, California, and to TRW Federal Systems Division, Redondo Beach, California.

NASA's Langley Research Center at Hampton, Virginia, was given the responsibility for coordinated design analysis and systems engineering.

By the end of 1985, as NASA was midway through the 21-months definition and preliminary design phase of development, the agency announced a change in the earlier conceptual phase. The earlier phase (A) had envisioned a central spine 400 feet long to which the modules and other station elements would be attached. The spine was called the Power Tower because the winglike photovoltaic solar panels that transformed sunlight into electricity were attached to it.

The Power Tower configuration was criticized by engineers as potentially unstable when forces

Figure 2.2. In this sketch, a more detailed rendition of the reference configuration is shown. Late in 1984, the configuration became known as the "Power Tower" because the solar power arrays were clustered at the top. Modules and docking facility were clustered at the base to provide gravitational gradient stability. The vertical truss is 400 feet long.

were applied to it, such as the docking of a shuttle orbiter. Vibrations and bending that would disturb pointing accuracy required by some scientific instruments could be expected, critics said.

The "Power Tower" was eliminated at the start of Phase B, the first of a sequence of major design changes that were to be made as the station evolved. The single vertical spine was replaced by two vertical trusses, the bridgelike structures that became known as "keels." The two keels joined at the top and bottom formed two sides of a huge rectangle initially 297 feet long and 126 ft wide. Across this framework lay a single horizontal truss 405 ft long. It supported the big solar panels that were emplaced at each end of the truss.

The station's four pressurized modules were clustered in a figure 8 near the center of gravity on the horizontal truss. As initially designed, each module was a cylinder 43.7 ft long and 14 ft in diameter, so that it could fit into the shuttle cargo bay and be carried up to and attached in orbit. The modules were interconnected at the ends by external nodes and tunnels with external airlocks.

A closed environmental control system was designed for the pressurized modules. It provided for the purification and reuse of air and water. The system would provide a normal sea-level-pressure atmosphere and water for drinking, food preparation, and bathing. With the closed environmental control system, only food and nitrogen would be resupplied to the station. The designs for this baseline configuration were to be completed by March 1986 and construction contracts were to be awarded in 1987. The station directorate projected that the station would be assembled in orbit and ready for occupancy in 1992. That prospect vanished January 28, 1986, when the space shuttle *Challenger* exploded during launch and its crew cabin plunged into the Atlantic Ocean. As a result of the investigation that followed, the top command of NASA was replaced and space station construction management was transferred from the Johnson Space Center, Houston, to headquarters in Washington. President Reagan recalled James C. Fletcher, who had been NASA administrator in the Nixon administration during most of the shuttle development period to head the reorganized space agency.

Fletcher named Andrew J. Stofan, director of the Lewis Research Center, to take charge of space station development as an associate administrator, June 30, 1986.

International participation in the space station, partly as a political and partly as an economic benefit, had been advocated long before it was authorized. It immediately became a reality, although fraught from time to time with policy disputes about the prospects of the station's militarization. Ultimately, NASA made agreements with Canada, the European Space Agency, and Japan to contribute elements to the station complex.

Building on its Remote Manipulator System technology, Canada would provide a Mobile Servicing Facility for satellites, open platforms, and possibly an auxiliary service vehicle. The European Space Agency offered development of a pressurized manned laboratory module called Columbus, free-flying experiment platforms in polar and low-inclination orbits, and a resources module. Japan proposed a multipurpose module, the Japanese Experiment Module, with a pressurized workshop and an exposed work deck. The ESA *Columbus* module would be connected initially with the station complex but would be outfitted to fly independently with a support module. ESA insisted that it remain in full control of its man-tended free flyer and its unmanned, polar-orbiting experiment platform.* In November 1987 the Council of the European Space Agency confirmed its interest in participating with the United States in the space station, provided its control of the free-flying module and the polar platform was set forth in the agreement between NASA and ESA. In addition, ESA wanted a provision for settling disputes arising out of the implementation of agreements. The question of potential military use of the space station by the United States disturbed the ESA ministers, who regarded as a problem the U.S. insistence on its right to determine the nature of peaceful uses of station elements.†

*The term "man tended" refers to intermittent visits by crew rather than occupancy.
†ESA announced March 18, 1988, that its negotiations with NASA on a bilateral memorandum of understanding with NASA on cooperative development and use of the space station were completed, a step toward a multilateral agreement among four "partners"—NASA, ESA, Canada, and Japan.

NASA formed an Advanced Technology Advisory Committee to study the application of automation and robotics to the station. This was of concern to some members of the congressional space committees who wanted to reduce the time that astronauts would be required to work outside of the pressurized modules. One focus of the committee's research was the development of a flight telerobotic system that would operate the Canadian Remote Manipulator and Mobile Servicing systems and free astronauts from other servicing jobs.

NASA also signed a memorandum of understanding with a private contractor, Space Industries, Inc., of Houston, for the exchange of information during Phase B development. The firm planned to fly a private pressurized laboratory that would be launched by the shuttle well in advance of the NASA station.

The laboratory, called the Industrial Space Facility, would be manned intermittently (man tended). That meant it would be visited from time to time by astronauts, to monitor or adjust industrial experiments. Plans called for the module to be built in Pittsburgh by the Westinghouse Electric Company, a partner in Space Industries, and assembled in Florida by Astrotech Space Operations, Titusville, in which Wespace, a Westinghouse subsidiary, owned majority interest. In order to get started, the Industrial Space Facility requested a subsidy of $700 million over five years from NASA in the form of advance leases of experiment facilities. This request was supported by both the White House and some members of Congress as a means of launching private enterprise into space.* The proposal called for NASA to launch two Industrial Space Facility modules in the early 1990s, two or three years before the national space station, and to defer payment until the Industrial Space Facility modules generated income from users other than NASA. NASA officials protested that the subsidy which would have to be paid out of the NASA appropria-

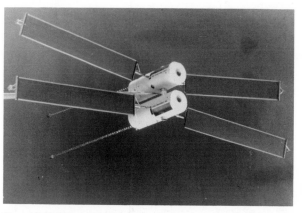

Figure 2.3. Space Industries, Inc., Houston, proposed this model of its proposed Industrial Space Facility, a man-tended, private enterprise station which NASA would subsidize by leasing services. The two center modules would be habitable. Solar wings would provide power. (NASA)

---

*Representatives Edward Boland (D–Mass.) and Bill Green (R–N.Y.) and Senator William Proxmire (D–Wis.) were reported as pressuring NASA to accept the lease subsidy by threatening to withhold some station funding. Boland was chairman and Green was ranking minority member of the House appropriations subcommittee on HUD and Independent Agencies that dealt with NASA funding, and Proxmire chaired that subcommittee in the Senate.

tion would seriously deplete funding for the construction of the NASA space station.

In March 1986, NASA accepted a firm proposal by the Canada Ministry of State for Science and Technology to do the preliminary design of the Mobile Servicing Center (or facility). The center would be equipped with remote manipulator arms, like those on the shuttle, and would be available to perform some of the space station assembly work. The arrangement was announced formally March 21, 1986, by Canadian Prime Minister Brian Mulrooney during a visit with President Reagan in Washington.

A space station development prospectus was presented to the Senate Subcommittee on Science, Technology, and Space April 23, 1986, by John B. Hodge, deputy associate administrator for the station.* Hardware development was to be started in

---

*Hodge served as acting associate administrator after Philip E. Culbertson, who headed the space station program since 1984 resigned in 1986. Hodge then became Stofan's deputy when Stofan was appointed associate administrator for the program.

the 1987 fiscal year and orbital capability would be achieved in 1994, Hodge said. "Our definition analysis confirms that such a date is realistic," he said "It also confirms our belief that a useful space station can be developed with the constrained budgetary outlook facing both NASA and the nation."

NASA officials were on safe ground as long as their forecast was confined to the progress in bending metal. However, the rate of development was not determined by technology but by money. Like the shuttle 15 years earlier, the space station was hostage to the budget. This time there was no war in Southeast Asia to compete with space, but there was the internal pressure of the budget deficit. Now space was competing for funds more than ever with the entitlement programs (Medicare, Medicaid, even Social Security), with farm subsidies, veterans benefits, and public housing programs.

Politically, the tendency was strong to view the space station as an option for the present, rather than as a necessity for the future. Those who feared

Figure 2.4. A docking port and Canada's Remote Manipulator and Servicing Systems are depicted here. The systems have been planned for use during both the construction and the operation of the Freedom Station. (NASA)

that politicians would yield to the deficit sought a strategy that would preserve station funding. One tactic, adopted by the administration, was to devise a policy that would entice free enterprise investment in space research and development. In this effort the space station was the key, for without it, Western Europe and Japan would be forced to deal with the Soviet Union. The U.S.S.R. had promised to provide commercial satellite launch service and offered limited experiment space on the *Mir* space station.

Government encouragement of free enterprise participation in space station development threatened to become counterproductive from the official NASA viewpoint when it resulted in subtracting dollars from national station development and lent them for an indefinite period to a small, private station, the Industrial Space Facility. NASA planners feared that if the Industrial Space Facility failed to attract users early in the 1990s, it would cast doubt about continuing a much larger investment in the national station later in the decade. This was a dilemma NASA faced so long as the government tried to sell the national station as a potential economic (as well as scientific) resource for the West. That ploy had already been tried with the shuttle and had failed, climactically, with the *Challenger* disaster.

Would Congress put up funds for the station simply as an orbital research and observatory facility, with the potential of becoming a base for lunar and interplanetary missions? Would the richest nation in human history invest a sum equal to only a fraction of its military budget to establish the human presence beyond Earth in the Solar System? These were questions that would test the vision of a whole generation of Americans, but they were not asked with any force during election campaigns of 1988, when the development of the U.S. national space station eluded definition as a public issue.

NASA had requested $420 million for the 1987 fiscal year to complete the definition and preliminary design of the station. The request included $15 million for utilization, $83 million for advanced development, $88 million for program management and integration, $17 million for operational readiness, $57 million for system definition, and $150 million for system development.

Hodge alluded to a directive by Congress that NASA complete the design for a Flight Telerobotic System for servicing spacecraft brought to the station. The allusion cited earlier hearings, when certain congressmen insisted that the station have such a facility to reduce the exposure of astronauts to extravehicular activity. NASA had agreed to seek development of such an invention.* The Flight Telerobotic System became an integral part of new space station technology. Hodge said that such a system would be the nucleus of a "smart front end" of an Orbital Maneuvering Vehicle, also to be invented, that would be used for precision servicing of free-flying platforms.

Hodge explained that NASA was planning to use a new solar dynamic power system to supply electrical power in addition to conventional photovoltaic panels. Solar dynamic power is produced by a dynamo turned by a sun-heated working fluid in place of steam. Its proponents claim that it is more efficient than photovoltaic (solar) cells, which are pieces of ceramic material (silicon, usually) that convert sunshine directly into electrical current. Adding solar dynamic generators would enable station designers to reduce the size of the solar panels, which create drag on a space vehicle in low Earth orbit, where there is always residual attenuated atmosphere. Drag would cause the station to lose altitude, thereby requiring the crew to fire its main engines, using limited propellant, to boost the big station to a safe altitude.

ACCESS AND EASE

The national space station authorized by President Reagan would be the first large structure to be built in space. In contrast to the 100-ton *Skylab* station of 1973, that had been launched by the mighty Saturn 5, the "Reagan" station would be assembled in orbit by astronaut steeplejacks from parts brought up from the Kennedy Space Center

---

*This task subsequently was awarded to Grumman Space Systems, Bethpage, N.Y., and Martin Marietta, Denver. These firms would compete in designing the robot.

sequentially by the shuttle. Like *Skylab*, the Soviet Union's space stations, the *Salyut* series and *Mir*, had been orbited essentially complete and serviced by manned spacecraft. Later, automated vehicle servicing was developed by the Soviets, and *Mir* was enlarged by the addition of annexes to the core unit.

With the abandonment of the Saturn 5, America's only heavy launch vehicle, the Reagan space station program depended entirely on the shuttle, at least at the outset. The configuration of the modules was defined by the size of the shuttle's 15 by 60 foot cargo bay. The shuttle thus acquired the role of a Procrustean Bed that defined the size and shape of modules not only for the space station but under a doctrine of commonality for the Moon and Mars as well.

The pace of station construction was limited to the three-shuttle fleet. The *Challenger* replacement orbiter would not be ready for flight until 1992, it was estimated. There was a weight constraint (about 16 tons) defined by the mass that the shuttle could lift to an altitude of 230 nautical (264.5 statute) miles on an eastward launch from Florida. The total mass of the space station was estimated initially at more than 200 tons. The extinct Saturn 5 could have lifted it in two flights. At least 12.5 shuttle payloads would be required to provide this mass, and hundreds of man-hours would be required to assemble it in orbit.

Building the space station would be the most demanding task for men, women and machines of the space age. The challenge was not merely technical but political. It would test whether a long-range goal of human expansion would be rationalized as a national endeavor in a period of socioeconomic stress.

As a practical consideration, there was a question of whether astronauts could perform station assembly tasks within the limits of time and manpower. Construction was planned at a lower orbital altitude than that in which the station would eventually be placed, so that the shuttle could lift the maximum payload per mission. The lower construction orbit posed the risk of early orbit decay and premature reentry. Manpower was limited by

the size of the shuttle crew and the time the labor force could spend in extravehicular activity.

Astronaut construction experience in orbit was limited. The first experimental effort to determine how well astronauts would perform tasks in EVA was made in 1966 in the Gemini program. The results, which showed that the men tired quickly unless they were well trained, demonstrated the necessity for practice in the neutral buoyancy water tanks at the Marshall Space Flight center and for precise task procedures.

Extravehicular activity was developed further as a specific skill during the Apollo program and low-gravity EVA was successfully carried out on the Moon. A one-sixth g field presented no difficulties once astronauts became used to it. Orbital free fall was another problem. There was no place to stand, and crews working outside space vehicles had to anchor themselves to fixtures with foot restraints and tethers.

In 1973 the *Skylab* crews made jury-rig types of repairs to the damaged station, demonstrating for the first time that this could be done safely and accurately. During 1984, the crew of *Challenger 41B* performed a spectacular job of capturing, repairing, and redeploying in orbit the Solar Maximum Mission observatory in April. Then in November of that year the crew of *Discovery 51A* retrieved two communications satellites, *Palapa B-2* and *Westar 6*, which had become stalled in low Earth orbit when their upper-stage rockets failed to deliver them to geostationary orbit.

In 1985 NASA designed two experiments in orbital construction techniques. These called for the assembly of erectable structures from the cargo bay of *Atlantis 61B* during the mission which flew November 26–December 3. One experiment, called ACCESS, was designed to prove an Assembly Concept for Construction of Erectable Space Structures. Another experiment, EASE, was designed to test the Experimental Assembly of Structures in EVA.

The experiments were performed during two EVA sessions November 29 and December 1, 1985 by Mission Specialists Jerry L. Ross and Sherwood C. Spring. They were assisted by Mission Specialist

Mary L. Cleave, who operated the Canadian Remote Manipulator System arm, and the mission pilot, Bryan D. O'Connor, Brewster H. Shaw, Jr., commanded the flight, which also deployed three communications satellites.

The results of the tests clearly showed that it would be feasible for trained crew persons to erect space station trusses and other structural components from the orbiter cargo bay. The tests took

two EVA sessions totaling 12 hours and 14 minutes.

On the mission's fourth day, Ross and Spring put on their space suits, crawled into the mid-deck airlock after breathing preparation, and exited the crew compartment through the aft bulkhead hatch into the open cargo bay. Because *Atlantis* was flying upside down, with the cargo bay doors open toward the ground, they perceived the Earth as above them,

Figure 2.5. Ultimate relaxation might be achieved by orbiting the Earth in the "easy chair" shown here. NASA calls it the MMU (Manned Maneuvering Unit). Strapped in it, an astronaut can fly freely in space as long as the fuel for his gas thrusters holds out. He is a one-man space ship. MMUs will be used to construct the space station (NASA).

Figure 2.6. The space station truss structure will be assembled piece by piece by astronauts, some in MMUs, some tethered to the orbiter cargo bay. In this photo, Astronaut Jerry L. Ross assembles a truss. He is supported by a foot restraint just below him at the end of the orbiter's remote manipulator system. This photo was taken by his partner, Astronaut Sherwood C. Spring. The truss assembly task was tested aboard the orbiter Atlantis on mission 61B November 29–December 1, 1985. (NASA).

the panorama of lands and seas passing by "overhead." (In orbital free fall, up and down are determined visually.) Ross and Spring snapped their tethers to slide wires on the orbiter's sill and took their positions at the Work Site, which was set up near the rear bulkhead hatch. Engineers had installed a device called a Mission Peculiar Equipment Support Structure. It was a beam extending across the 15-ft diameter of the cargo bay. Attached to it were foot restraints that enabled the astronauts to maintain a firm position at the lower and upper levels of the "mission peculiar" support structure.

The object of the ACCESS experiment was to erect a 45-foot truss with a triangular cross section. First step was to clamp an 11-ft assembly fixture horizontally to the support structure. The fixture consisted of a central tubular mast with guide rails on which the three-sided truss was assembled. The assembly fixture, mounted on a pivot, was rotated on the support structure to a vertical position and locked. The three guide rails were then opened out to form the assembly structure. Raising the assembly fixture into position took 3.5 minutes.

The astronauts secured themselves into foot restraints, one at the side of the assembly fixture and one elevated above it. They were ready to put the truss together. It consisted of 93 tubular aluminum struts one inch in diameter. The vertical struts were called longerons. The horizontal struts were called battens. Each was 4.5 feet long. They were braced by diagonal struts 6.36 ft long.

The struts were connected at each end to joints called nodes. Each node was attached to a guide rail on the assembly fixture. One end of a strut was fitted into a spring-loaded sleeve in the node and locked into place. The locking fixture on the sleeve was cocked initially so that it would snap closed when the strut end was inserted. That would allow the astronaut to insert the strut with one hand. However, the astronauts reported that they had to lock the struts into the nodal sleeves manually for the most part.

As defined by the assembly fixture guide rails, the truss would have a triangular cross section consisting of 10 sections or bays. The nodes and struts had been stowed in three canisters at the work station.

The construction plan called for the crew to maintain their positions on the assembly fixture for a time and then switch (to give relief to the man in the lower position, who had to handle more equipment). One astronaut worked with the upper joints; the other, with the lower joints. The assembly fixture was manually rotated about its long axis, to allow the astronauts to make the joint connections without leaving their foot restraints. When a bay was completed, it was pushed up the assembly fixture rails and the next bay was started. The fixture accommodated two bays at a time. Ross and Spring exchanged stations after assembling five bays.

The 45 foot truss was completed in 25.6 minutes. Bedecked with a U.S. flag, it was photographed and then disassembled in 18.2 minutes. The astronauts lowered the assembly fixture and locked it horizontally on the support structure in 4.7 minutes. The entire ACCESS operation had taken 53.32 minutes of an allowable time line of 105 minutes. Ross and Spring reported that the truss felt rigid and stable when they removed it from the assembly fixture and moved it about by hand. Ross said that the truss, with a mass of 190 pounds was easy to manipulate manually from either end or the middle. Ross and Spring had rehearsed the ACCESS experiment before the flight for 12 hours in the neutral buoyancy tank at Marshall and in a similar tank at the Johnson Space Center. The practice had paid off, allowing them to do the task in about one-half the allowable time.

After a short break, Ross and Spring undertook the demonstration of the Experimental Assembly of Structure in EVA (EASE). The task was to build a large tetrahedron, disassemble it, and repeat the task seven times. The purpose of the repetition was to show if performance improved with practice and whether the astronaut would experience less fatigue with more experience. On this test both astronauts had to work floating freely most of the time. Only one of the connector clusters on which the four-side structure was formed could be reached from the work station foot restraint. Ross was the

Figure 2.7. Astronaut Sherwood C. Spring checks the joints on the 45-foot long truss assembly after taking Jerry Ross' position on the remote manipulator system's foot restraint. The flag is attached to signify that the job is done. (NASA).

Figure 2.8. The second part of the assembly test on Atlantis 61B was called EASE (Experimental Assembly of Structures in Extravehicular Activity). The experiment called for erecting a tetrahedron of 12 foot poles. Ross works with the assembly at the left, clinging to a pole. Below in the cargo bay is the cover for the Mexican communications satellite, Morelos, which was deployed on this mission.

low man and Spring the high man during the first four assembly–disassembly cycles. They then exchanged positions. Metabolic rate tests showed that the low man worked harder, as on the ACCESS test.

The hardware consisted of six 12-foot beams, three removable connector clusters, and one cluster of connectors fixed to the work station. The low man connected two vertical riser beams to the base cluster and a node to the equipment support structure. After that was done, both astronauts moved to the top of the EASE structure to attach a horizontal cross beam to join two risers. The low man then installed a third riser beam and the high man added two cross beams to join the three risers. The structure took on the shape of the frame of a four-sided kite, with a mass of 395 pounds.

The first four assembly–disassembly cycles were completed in one hour and seven minutes of a two-hour time allocation. The final four cycles were done more slowly. Both crewman said that the work was tiring for the free-floating astronaut because he had to hold a correct body position while he used his hands and forearms to turn the beams into place. On the mission's sixth day, Ross and Spring performed their second EVA to test construction of the truss and tetrahedron from the 50-foot Remote Manipulator System arm. Astronaut Mary Cleave operated the arm from the orbiter's aft flight deck. The test was designed to show how well the work could be done from a movable platform.

The astronauts reported that they found it easier to assemble the EASE tetrahedron from the foot restraint fixture on the arm than to do it while floating free. It was something to stand on. But the ACCESS truss bays would take longer to construct because only one man could work on assembly, and he had to wait while the arm moved him from one work site to another. Consequently, after the first nine bays of the truss were assembled from the fixed work station, Ross climbed into the Remote Manipulator System arm foot restraint and was hoisted to the top of the truss to complete the tenth bay. He also simulated the installation of a power cable on the tower from base to top as Mary Cleave moved the arm to lift him four feet at a time up the entire 45 foot length of the truss, so that he could clip the "cable" (a nylon rope) to each bay. Ross next detached the rope and moved the entire truss by hand away from the assembly fixture and held it. He reported that it was easy to lift and control (in microgravity). He then reattached the truss to the assembly fixture. Spring next took a turn performing these maneuvers.

The ACCESS and EASE experiments had shown the feasibility of constructing the initial space station's transverse boom 360.89 feet long by similar techniques from the shuttle orbiter cargo bay. At least it seemed feasible to do it in 45 foot increments or less.

The ACCESS truss design, the material, and the nodal joints were later modified, but no further opportunity came to test the modifications until the shuttle resumed flight in 1988.

THE MAN-TENDED STATION

In view of the reduced shuttle fleet, Congress directed NASA to study the effect on the nation's space objectives of part-time occupation of the space station as a cost-saving alternative to full-time occupation. Part-time use was referred to as the "man-tended" option. It would save money by reducing shuttle flights for resupply. *Skylab,* for example, had been man-tended. NASA rejected the proposal but nevertheless was required to justify its position. The agency dutifully made a study and reported that man-tending would not save money in the long run and would adversely affect materials-processing experiments and life sciences research. Of 37 materials-processing experiments studied, 33 could not be done in a station that was not continuously manned, nor could 33 out of 54 technology development missions. Of 138 missions, only 39 could be completed without alteration in a man-tended operation and 93 could not.

The study concluded that although the man-tended option is technically feasible and could carry out some useful functions, it would force postponement of experiments important to the advancement of manned space flight and increase the cost of developing the permanently manned station, the

Figure 2.9. Astronaut Spring takes a turn connecting the EASE poles.

national objective. Members of Congress who wanted NASA to consider the man-tended option did not press it further. A government man-tended station would duplicate the Industrial Space Facility, which also was planned as a man-tended facility.

As national station development evolved slowly through the Phase B period in mid-1986, the first manned mission aboard the new Soviet *Mir* space station was completed successfully July 16. Leonid Kizim and Vladimir Solovyev landed after 125 days in orbit. The journal *Aviation Week and Space Technology* prodded the government by observing that Soviet cosmonauts had accrued 104,374 hours of orbital flight compared with 42,453 by American astronauts.

The Dual-Keel Station. NASA published a full description of a "dual-keel" space station it planned to build at a cost of $8 billion and discussed it widely in June 1986. Station assembly was slated to begin in the first quarter of 1993 and was to be completed in 1995. It would be man-tended during the early part of the construction period, but by 1994 it would be capable of being permanently occupied. The station would comprise habitation, laboratory, logistics, and international modules interconnected by nodes, airlocks, and tunnels. Auxiliary and support equipment attached to the main trusses, including a servicing facility and the telerobitic servicer required by Congress. There would be accommodations for attaching payloads and a berth for the shuttle orbiter. The system included ESA platforms co-orbiting with the station and in polar orbit and a U.S. polar-orbiting platform. There would be provision for storage, for utilities, and for docking and servicing orbital maneuvering and orbital transfer vehicles. Electric power would be generated by photovoltaic cells and solar dynamic turbines. The station would have its own propulsion system to maintain or extend its orbit. These features were described in some detail by the space agency's *Systems Requirements Review*, published by the station office. By mid-1986 it appeared that a final plan had evolved.

The dual-keel design was so called because it featured a pair of vertical trusses, each 310 feet

long joined horizontally by 150-foot trusses at each end to form a metal rectangle, a huge picture frame hanging over the Earth. Running horizontally across the middle of the rectangle would be a central truss to which the habitat, laboratory, logistics, and international modules would be attached.

The dual-keel configuration was the second to be considered after the station was authorized in 1984. It succeeded an earlier design consisting mainly of a 400-foot-long "power tower," so called because photovoltaic arrays blossomed at the top of it. The stability of the tower was questioned, although in theory it was anchored by gravity gradient. The dual-keel design was replaced in 1987 by the "revised" single-truss design to cut costs. Crewmen, regardless of nationality, would live in the U.S. habitat module.

The U.S. habitat and laboratory modules would be 44.5 feet long and 14.6 feet in outside diameter to fit in the shuttle cargo bay. The inside diameter of the modules would be 13.8 feet. Modules would be pressurized with air to 14.7 pounds per square inch.

The 1986 *Systems Requirements Review* described the basic construction process: Astronauts would assemble trusses with a 16.4 foot cross section to form the perimeter of the picture frame support structure. Halfway between the ends of the 310 foot keels they would erect a transverse truss across the keels and extending beyond them to support the solar arrays. The modules would be anchored to the transverse beam and interconnected by nodes at the ends.

The *Review* specified that construction crews would wear the shuttle space suit for EVA during the first phase of construction. NASA was exploring the cost of developing a higher-pressure suit that would require less breathing adaptation than the shuttle suit.*

Other elements of the station included the Logistics Module, for pressurized and unpressurized cargo, propellant, and fluids; a servicing facility, to

---

*Derived from the Apollo space suit, the shuttle suit holds pressure of 4 psi. To avoid the bends, cabin pressure (14.7 psi) had to be dropped to 10.2 psi for 24 hours before the EVA started and crewmen had to breathe pure oxygen to purge nitrogen from the blood. It was planned that the new suit would maintain a pressure of 8.4 psi and be more flexible than the shuttle suit.

permit crew access to service payloads and free flyers; the Telerobotic Servicer, for station assembly, maintenance, and payload servicing; the Canadian Mobile Servicing Center with manipulator arms that would travel along a truss on a (NASA-supplied) transporter; and the docking station for the shuttle.

Modules provided by the United States would have a common structure so that they could be interchangeable at the station or be used in lunar orbit or as surface shelters. Engineers expressed the view that the entire station could be adapted as an interplanetary transport with addition of powerful rocket engines and propellant supplies. Commonality had become a watchword in future development planning. There was a widespread belief among NASA planners and engineers that the technology and hardware developed for the space station would be usable for lunar and planetary missions without extensive modification.

The European and Japanese modules, however, could not be expected to fit into this scheme. Their configurations would differ from that of the U.S. modules to meet the requirements of their sponsors. ESA's module, *Columbus*, was built as a multipurpose laboratory for international use. It would be equipped for research and experiments in fluid physics, life sciences, and materials research. It would consist of four *Spacelab* segments 10 inches less in diameter than the U.S. modules. Only the experiment equipment racks would be standardized to match those of the U.S. laboratory.

Figure 2.10. Configuration of the arrangement of space station modules circa 1986. Connecting nodes were later redesigned. (NASA).

Japan's Experiment Module (JEM) was to consist of pressurized and unpressurized sections, a scientific equipment airlock, a remote manipulator, and a logistics module. The JEM was to be 13.78 feet in diameter. It would be designed to be returned to Earth to be refitted, and reflown. JEM was to be compatible for launch by an expendable rocket, the H-20, which the Japanese Space Agency planned to fly in the mid-1990s.

Assembly and Construction. Station assembly was planned initially at an altitude of 220 nautical (253 statute) miles to enable the shuttle to reach the assembly site with heavier payloads than it could lift to higher altitudes. The station, however, would operate at 250 nautical (287.5 statute) miles at the standard shuttle flight inclination of 28.5°. This arrangement would bring the station over the space center and simplify rendezvous and docking for shuttle crews.

The *Systems Requirements Review* reported that 14 shuttle flights would be required to complete station assembly. If the assembly began in January 1993, as planned in 1986, all elements would be in place by 1995, according to the *Review*.

The modules were the largest components the shuttle would be required to lift to the construction orbit. They were formed of a pressurized shell, a single, waffle-patterned, cylindrical structure enclosed and insulated by an outer aluminum shell which was designed as a bumper shield against meteorites. These ranged in size from tiny particles of dust to ball bearings. Between the shielding and the pressure shell were layers of mylar, kapton, and kevlar insulation. At each end of the module there would be a hatch 50 inches square. Attached to it would be a joining node, a sphere 10.5 feet in diameter with six 78-inch ports for the attachment of tunnels. The nodes and pressurized tunnels would provide access to other modules and would allow the module to be expanded in three dimensions.

The tunnels would be aluminum cylinders 78 inches in diameter and 19.8 feet long. They would be pressurized but otherwise, the *Review* said, "austere."

Like a true spacecraft, the station would be equipped with a guidance and navigation system.

It would enable the crew to maintain or adjust orbital altitude. Gas thrusters would maintain station attitude (position) and height would be stabilized by continuous firing of microthrusters or, alternatively, by firing orbital maneuvering or space engines about every 90 days to regain altitude lost by friction with the residual atmosphere, even at 287.5 miles.

To meet the shuttle on resupply missions, the station would drop in altitude to 253 miles. It would regain cruise altitude by firing its orbital maneuvering engines, to be located at the four corners of the frame. Hydrogen–oxygen or hydrazine and nitrous oxide have been considered as propellants.

The initial electric power allotment for the station was set at 75 kilowatts, enough electricity to run 25 average homes, the Review said. About 50 kilowatts would be allocated to experimenters and 25 for housekeeping. The free-flying platforms would have their own power systems, ranging from 8 to 12 kilowatts. A hybrid power-generating system was envisioned. About 25 kilowatts would be produced by photovoltaic cells from sunlight and 50 kilowatts by a sun-heated turbine generator. Designers liked the solar dynamic heat engine system better than the photovoltaic system because the huge panels of solar cells in the latter increased the drag on the station and accelerated its loss of altitude.

The solar power system would deploy four arrays of photovoltaic cells 33.5 by 43.6 feet. The power could be distributed directly and some of it stored in nickel–hydrogen batteries. For the solar dynamic system, the station would erect parabolic mirror segments to focus sunshine on a working fluid, heating it to 2000 degrees Fahrenheit to drive the turbine generators. The mirror arrays would be one fourth the size of the photovoltaic solar panels.

If photovoltaic power were used alone, generating 75 kilowatts would require eight solar arrays, each 30 by 80 feet. Solar cells were viewed as inefficient; *Review* data indicated that they could convert only 8 percent of the sunshine they received into electricity. Energy generation by this system stops when the spacecraft passes into the "night side," or umbra, of the planet. Solar dynamic power systems keep on working during umbra passage by storing heat in molten salt during the approach to umbra, so that the heat radiates and keeps the turbine working as the spacecraft passes through umbra.

Communications and tracking systems on the station would use the S-band radio frequency transmission through the Tracking and Data Relay Satellites (TDRS) during station construction. Once in orbit, the station's principal mode of communication would be on the Ku band radio frequency which can carry video signals as well as engineering and scientific data in large volume. The station would be equipped with 22 television cameras. Tracking would be done through a Global Positioning Satellite. Station navigation would use star sightings.

The most visible part of the station would be its dual-keel frame, the size of a football field. The erectable truss lattice would consist of composite graphite tubes connected by aluminum fittings. The tubing would be 2 inches in diameter and could be grasped by an astronaut wearing gloves. Spring-loaded connectors used in the ACCESS and EASE experiments were rejected in favor of simpler fittings devised at the Langley Research Center.

The question of how a crew could escape from the station in case of an accident worried some members of Congress and was studied by the *Systems Requirements Review*. Reviewers suggested that a lifeboat could be included in the space station budget. It would be in addition to "safe havens"—areas of the station that could be independently pressurized and sealed off against leaks, like water-tight compartments in a ship, in case of damage to the station.

The 1986 *Systems Requirements Review* noted that the *Challenger* accident had raised serious questions about the readiness of the shuttle as a rescue vehicle. Unlike *Skylab*, where the *Apollo* spacecraft was docked while the crew was on board, the space station of the mid-1990s was planned initially without any means of escape in case of an accident between shuttle resupply visits.

Retrenchment. By fall 1986, it was becoming evident that the dual-keel space station could not be built for its earlier estimated cost of $8 billion (1984 dollars). Administrator Fletcher ordered a program

review and new cost study. Costs were refigured by the centers assigned Phase B work packages. A new estimate appeared. It was $14.5 billion (1984), nearly double the one made hardly a year earlier. This calculation generated shock waves through the administration and Congress. It did not include the cost of experiments, transportation, personnel, facilities and operations.

Fletcher presented the revised estimate to senior officials at the White House in February 1987. It was clear by then that if the Reagan administration wanted to go ahead with the space station, NASA would have to repeat the compromise it had made 15 years earlier on the cost of building the shuttle. It would have to cut back to a cheaper station.*

Throughout February and March 1987 the space station directorate looked at options for reducing cost. By April 1987 senior engineers reached a decision on a revised baseline configuration. It eliminated the dual keel and a number of service features and projected a single transverse boom as the anchor structure to which the pressurized American, European, and Japanese modules would be attached. At each end of the boom, photovoltaic arrays would be mounted to generate a total of 50 kilowatts of electric power. Two external payload attachment points would be provided on the boom. The revised station budget would include the Flight Telerobotic Servicer and an unmanned, free-flying, polar orbiting platform, but not the Orbital Maneuvering Vehicle.

The U.S. laboratory and habitat units and the ESA and Japanese modules would be equipped as in the dual-keel design. Their architecture would be unchanged except for larger nodes and tunnels joining them. These fixtures would accommodate the avionics that had been designated for attachment to a truss on the dual-keel station. Instead of being merely passageways for crewpeople to move from one module to another, the nodes and tunnels would be fitted with equipment to achieve a more compact utilization of pressurized station volume than that planned in the dual-keel design.

* In 1972 an original plan to develop a fully reusable shuttle with a winged booster capable of flying back to the launch site was replaced by a partly reusable shuttle with a throwaway external propellant tank and solid rocket boosters.

Figure 2.11. Revised Freedom Space Station architectural plan as submitted to the U.S. House of Representatives Committee on Science, Space and Technology in November 1987. The dual keel had been removed—for the time being. (NASA)

With the enlarged nodes and tunnels, the revised baseline configuration would provide 31,000 cubic feet (878 cubic meters) of usable, pressurized volume—more than two and one half times that of *Skylab*). The revised station would provide all the essential facilities of the dual-keel design.

It did not reduce the total pressurized volume planned for the dual-keel station, but it provided less extensive industrial research and satellite repair facilities.

The revised baseline configuration was designated as Phase I or Block I of the space station. After it was on orbit, it could be enlarged by adding the dual-keel trusses with their facilities. This would constitute the Phase II or Block II station. At a Washington briefing April 6, 1987, Andrew Stofan, associate administrator for the station, asserted that the only way that the station would be started under budget restraints was to defer some planned capabilities and structures, to reduce cost.

In the Block II station, electric power would be increased by adding a solar dynamic generator. Satellite servicing bays would be added. Other additions would be a co-orbiting (U.S.) unmanned platform, the Orbital Maneuvering Vehicle and possibly the long-planned Orbital Transfer Vehicle. The OTV could boost satellites to geostationary orbit (22,300 miles above the equator) and send payloads to the Moon.

A persuasive factor in reducing the mass of the station was the agency's concern that the three orbiter shuttle fleet did not have the capacity to build the 1986 station on the projected time line, even though the fourth orbiter had been authorized in 1986 to replace *Challenger* and Congress had appropriated $2.1 billion for it in NASA's 1987 fiscal year budget. Rockwell International was awarded a $1.3 billion cost plus award fee contract to manufacture it. The balance of the appropriation would be spent to replace the main engines, space suits, and other equipment lost with *Challenger*.

The fourth orbiter (OV 5) would not be ready to fly until 1992. Until then three shuttles had to provide the construction lift for the station. The acting program manager for the station, John W. Aaron, was reported as estimating that 19, instead of 14, missions would have been required to build the dual keel station. At a flight rate of five a year, the construction period would have been four years. A more pessimistic estimate claimed that 31 launches would be required to put the dual-keel station up and that would take eight years.[1]

The revised baseline configuration or Block I station would have a price tag of $14 billion—almost as much as the dual-keel configuration it was to replace. The price covered $12.2 billion for development, $1.5 billion for an emergency rescue craft, and $300 million for the Flight Telerobotic Servicer that Congress wanted. The enhanced Block II station with the dual keel added would cost an additional $3.9 billion, including the Orbital Transfer Vehicle. The total of $17.9 billion was for research and development only. The cost would rise to $27.5 billion for "full development and deployment," according to NASA and National Research Council estimates. NASA then issued requests for proposals from industry for the final design and development stages of the station, with separate proposals for Block I and Block II.

In theory, phased construction that would enable the national space station to be completed and operating by the mid-1990s would keep costs to an acceptable level. This was the burden of the plan NASA submitted to the appropriations committees of the House and Senate. But members of these committees had plans of their own. The House Appropriations Committee obliged NASA to include four features in the construction and deployment of the station, as a condition of receiving an appropriation. These were (1) A fully equipped materials processing laboratory (in orbit) before the habitat module was launched, (2) a minimum of 37 kilowatts of electric power to be available on the station before a habitat module was launched, (3) the attachment of pressurized payloads also before the habitat module was launched; (4) accommodation for life sciences research, including essential animal facilities. Unless these conditions were met, the committee stated, it would not approve the release of the Request for Proposals (RFP) for the final design and development of the station; in effect, it would stop the project.

These demands were in addition to the earlier committee requirement for development of the

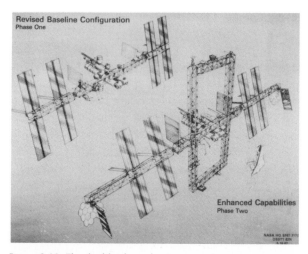

Figure 2.12. The dual keel was being retained as an option, to be added sometime later. This sketch shows the revised baseline configuration or Phase I and the enhanced capabilities (Phase II). The two concepts were actually drafted as of May 18, 1987, as dated lower right. (NASA)

Flight Telerobotic Servicers and another requirement for the early capability of man tended operations as part of the revised baseline configuration. The Appropriations Committee insisted on a "fair allocation of station resources" and required that operating costs would be shared by space station partners (Europe and Japan). Moreover, the committee insisted, no foreign module except the Canadian Mobile Serving Center was to be launched until all of the U.S. core infrastructure and full power were established on orbit.

NASA had no choice but to accept these demands. The extent of the political interference they represented was unprecedented in the space program. It went so far as to dictate the order of station construction and the use of specialized equipment to satisfy the concerns of private commercial and industrial interests. In the Apollo era, political intervention to this extent would have drawn fire from the media. Its unchallenged appearance now illustrated the degree to which NASA had lost prestige and public confidence. The spectacle of politicians telling engineers how to build a space vehicle seemed to be uniquely tolerable in the post-*Challenger* period of shuttle recovery.

ORDER OF BATTLE

In a report to the congressional appropriations committees, NASA had outlined the sequence in which the revised baseline station would be

Figure 2.13. Enhanced docking module drawn in this representation allows crew to enter space station while the orbiter cargo bay is being serviced by the Mobile Servicing Center and the Remote Maniplator System or "Canadarm." (NASA)

launched and assembled. A total of 16 station missions were planned: 11 major assembly flights, three logistics flights, and two missions to put up U.S. and ESA polar orbiters. The major assembly and logistics missions would be launched from the Kennedy Space Center in Florida, and the polar orbiting platforms would be launched from Vandenberg Air Force Base, California. NASA considered the possibility of launching the polar plat-

forms aboard Titan 4 expendable rockets instead of the shuttle. The 11 major assembly missions would require the shuttle as a construction base. The order was as follows:

MISSION 1. The photovoltaic power module, starboard transverse boom, aft starboard node, a control package, part of the reaction control system, antennas, and assembly support tools, including the Flight Telerobotic Servicer, jigs, and other tools. The station

Figure 2.14. Astronauts would be able to observe the orbiter docking with the station from a cupola as shown in this portion of a station mock-up. (McDonnell-Douglas).

Figure 2.15. This mass of shock absorber structures and wires is part of the "innards" of the docking mechanism developed by McDonnell Douglas Co. to allow the shuttle orbiter to dock gently with the space station. The docking of two space vehicles in orbital free fall was developed in Project Gemini and highly refined in the Apollo lunar landing program where precise docking of the Lunar Module with the Apollo Command Module was a life-and-death proposition. The shuttle orbiter-station docking is being designed to avoid shaking the station and disrupting delicate experiments and observatory instruments. (McDonnell Douglas).

would then be functional, with 25 kilowatts of power. With elements of the reaction control system, the structure would be capable of maintaining orbital altitude.

MISSION 2. This launch would support the port structure and systems, complete the horizontal boom, provide the full 50 kilowatts of power, and provide full guidance and navigation as well as control capability. It would include the control moment gyroscopes.

MISSION 3. It would deliver the first package of attached payloads, the heat rejection system, the airlock and docking adapter, the remainder of the reaction control systems, additional antennas, and phase one of the Mobile Servicing System.

MISSION 4. It would add a second remote manipulator system to the Mobile Servicing System, a hyperbaric airlock and reaction control system tankage. A payload would be attached to the transverse boom.

MISSION 5. The U.S. laboratory would be brought up to orbit and outfitted with nine of its 44 racks. Man-tended capability would be achieved.

LOGISTICS 1. On this mission the rest of the laboratory equipment would be delivered.

MISSION 6. The partly outfitted U.S. habitation module would be delivered.

MISSION 7. The balance of the habitat module would be delivered. The port and starboard nodes would be assembled with cupolas to aid in docking the shuttle and in operating the Mobile Servicing System. Extravehicular support equipment and part of the Mobile Servicing System maintenance depot also would be delivered.

POLAR 1. The U.S. polar-orbiting platform would be launched from Vandenberg, probably by an expendable rocket.

POLAR 2. The European Space Agency platform would be launched from Vandenberg, again by an expendable rocket.

MISSION 8. A crew of four astronauts would be flown to the station to commence permanent manned occupation, along with logistics modules and equipment to sustain the crew for six months. Four extravehicular space suits would be delivered on this flight.

MISSION 9. The Japanese Experiment Module and the first section of its open deck would be delivered.

LOGISTICS 2. Logistics equipment would be supplied and two astronauts would be added to the crew, making a total of six.

MISSION 10. The European Space Agency's pressurized laboratory would be delivered.

LOGISTICS 3. Additional logistics would be delivered and two more astronauts would join the crew, making a total of eight persons aboard the station. Station tours of duty would be extended to 180 days to reduce the number of shuttle resupply missions to four or five a year.

MISSION 11. The sixteenth and final flight in the assembly sequence would carry the second section of Japan's unpressurized deck and the Japanese Experiments Logistics Module.

THE NRC REVIEW

In the spring of 1987 senior White House officials and Administrator Fletcher asked the National Research Council (NRC) of the National Academy of Sciences and the National Academy of Engineering to review the space station program. The NRC formed a committee for this purpose under the chairmanship of Robert C. Seamans of the Massachusetts Institute of Technology, a former NASA official. The committee issued its final report September 10, 1987.

The report was addressed to the President's Assistant for National Security, Frank C. Carlucci; the director of the Office of Management and Budget, James C. Miller II, and the President's Science Adviser, William R. Graham. In his letter of transmittal, Seamans characterized the space station as the "most ambitious space project the nation has ever undertaken." That perception had not been widely realized; the space station had seemed secondary in scope to the Apollo manned lunar landings and more on the order of the shuttle. But the NRC committee viewed the challenge of the space station program to be as demanding as that of Apollo and its potential even greater as a base for future lunar and planetary exploration.

"It will require tens of billions of dollars over a period of several decades," the Seamans letter said, "and will absorb much of the energy of NASA for much of that period." Seamans warned that "it must have enduring support across administrations" and "this support must be generous enough to provide adequate hardware and testing to ensure that it can be operated safely for decades . . . and

an adequate space transportation system for its deployment, assembly and operation."

Seamans, however, had reservations about the state of the Space Transportation System. "The current space shuttle is barely adequate for the limited purpose of deploying the space station," his letter said, "and it is inadequate to meet broader national needs in space." In sum, the shuttle was "a major concern." As chairman of the review committee, Seamans recommended "in the strongest terms" that the shuttle be upgraded with new, improved solid rocket motors; that it be supplemented with expendable launch vehicles and that a heavy lift launch vehicle be developed for use in the latter half of the 1990s.

In its overview of the space station program, the committee depicted the station as a historic step, with widespread international as well as technological implications. The United States was committing itself to a major national project "that will of necessity require tens of billions of dollars and take a decade to bring to operational status," the review said. The station cannot be considered a "one administration program" nor can it be developed "on the cheap," the review warned.

It added the station will be viewed as a symbol of the competition in space between the United States and the Soviet Union; "moreover, our international partners, Canada, Japan and the nations of the European Space Agency, are relying on the space station as the foundation for their manned space flight program."

The committee concluded that the station could become "a visible demonstration of American technological and operational prowess in space" and the base "for assembly of the next generation of large, scientific instruments to observe the universe . . . a way station for manned lunar and planetary missions." The report characterized the station as a national goal that was an essential element of an aggressive manned space program. More specifically, it concluded that the early scientific and engineering uses of the station "are reasonably well understood." The Block I configuration was "a satisfactory starting point for the station." The design, the reviewers said, reflected "thoughtful compromises among the priorities and sometimes conflicting requirements of its early scientific and engineering users." The reviewers added that a commitment to a particular configuration for Block II would be premature in the absence of agreed long-term space objectives. "Indeed, the next phase of the space station could go in any of several directions," the committee said.

The committee found no operational and little scientific relationship between the polar platform included in Block I and the main station. Such platforms, it said, should be evaluated on their own performance. The committee added that prospective users of the co-orbiting platforms planned for Block II would be unlikely to benefit from man tending of the platforms, a proposal from platform users. "The administration should clarify its long term goals in space before committing the space station to a specific evolutionary path beyond Block I," the committee said.

Emergency Rescue. The committee was not optimistic about the shuttle, which it noted was the only means the nation has of deploying the space station. While this was not "infeasible," it said, deploying the space station with the post-*Challenger* shuttle "will be difficult and risky." The committee recommended that the post-*Challenger* shuttle have improved solid rocket motors and that NASA acquire an orbiter in addition to the replacement for *Challenger* before space station deployment was started.

In addition to restoring the shuttle fleet to four vehicles plus a spare, the committee advocated the development of a heavy lift launch vehicle for use in the latter half of the 1990s. It proposed also that expendable launchers be used for station logistical support to supplement the shuttle. The committee urged that a crew emergency rescue vehicle be made a mandatory requirement of station development and operation. NASA should consider also using a man-rated expendable launcher* as a back-up to the shuttle for manned access to the station or escape from it.

The National Research Council report essentially confirmed the engineering and scientific va-

lidity of the space station plan, with some limitation. The Block I station would provide a satisfactory materials and sciences laboratory but would not provide a complete life sciences facility nor separate it from materials sciences research facilities. It would be difficult to provide for large animals or to install a large, variable force centrifuge in the laboratory.† It may be necessary, said the report, to add a dedicated life sciences module if the United States plans to undertake long duration, manned interplanetary missions.

The Block II station, as designed in 1987, could not be used as a base for manned lunar or interplanetary flights without substantial modifications. Nor would it provide effective Earth observation. For that purpose an Earth observatory should be in a polar orbit, the committee said.

Insofar as Solar System experiments and general astronomy observation were contemplated, these programs could be deployed effectively on a free-flying platform. For that reason the Block II addition of the dual keel, with its perpendicular trusses, would not add much to the station's utility as a science platform. The committee added that "it is important that space sciences not be confined, made hostage if you will, to the space station and shuttle."

Falling Hardware. The committee warned that with the limitations of the shuttle fleet, NASA faces the risk of losing space station elements by orbital decay, allowing them to fall back into the atmosphere and burn. To avoid this, NASA's assembly program provided for the attachment of reaction control engines to the earliest hardware deployed in orbit. The engines would be fired by radio control to reboost the hardware into safe orbits when necessary. However, as the committee noted, the space agency has had to consider the possibility that reboost systems might fail.

If the reboost system failed in the deployment program NASA outlined in 1987, the orbit in which space station elements would have been placed would decay in only 20 days. The altitude of this orbit was about 253 miles, as mentioned earlier. It

---

* A launch vehicle reliable enough to launch a manned spacecraft.

† A centrifuge that would enable humans or animals to experience artificial gravity for periods of time while in orbit had been considered on an advanced Skylab station that was never built.

was selected on the basis of shuttle performance and maximum payload. The committee suggested that vulnerability to orbital decay could be lessened by reducing the payload weights of the first two launches and allowing the shuttle to place them in higher orbits. An extra launch could be added to the series to make up for the reduction in payloads of the first two launches.

Instead of NASA's program of 16 launches to build the station, the committee calculated that 18 launches would be necessary to deploy the Block I station and an extra, make-up launch would bring the total to 19. However, by improving shuttle performance with advanced solid rocket motors, the early packages could be boosted high enough without cutting their weight to increase orbital lifetime to 240 days. In the event of a reboost failure, there would be enough time to launch a shuttle to boost the station to a safe altitude.

An improved shuttle with greater lift capacity than that calculated for the existing fleet would make it possible to cut four flights from the 18 flight station deployment sequence and would reduce the time crews would have to spend in extravehicular activity by about 40 hours, the committee calculated.

Commenting on NASA's decision to increase crew stay time from 90 to 180 days in the station, the committee noted that although slower crew rotation would ease pressure on the shuttle program, its effect on the crew was uncertain. There was insufficient medical and psychological data in the American space program to provide "high confidence" that doubling stay time in orbit would not have adverse physical and psychological effects. However, the committee understood that NASA intended to increase crew stay time gradually and study the effects. There was also an expectation that the Russians would share their findings about long-term tours of cosmonauts in *Salyut* and *Mir* stations.

The NRC committee published its cost estimates for the Block I station as of June 30, 1987, for research and development only (see table 2.1).

In addition to research and development, costs of deploying the Block I station were taken into account by the committee. These included shuttle transportation, station operations prior to full operational capability, NASA personnel, related facilities, and shuttle modifications for the station, including an extended-duration orbiter. Thus, total costs for development plus deployment of the Block I station were listed in the committee's revised estimate as $21 to $25 billion (1984 dollars) and $25 to $29.9 billion (1988 dollars). The Block II enhanced configuration would cost an additional $6.5 billion in 1984 dollars. (The 1988 dollar equivalent was not published.)

These estimates were based entirely on NASA figures. It was not surprising that they approached the total costs of Project Apollo, for as the space station program developed in the second half of the 1980s, it was clear that its magnitude would exceed Project Apollo and the landings on the Moon.

Alternative Designs. The National Research Council Committee on the Space Station examined several alternative station configurations. Most of them were rejected for scientific and engineering reasons.[2] Five preliminary designs were reviewed by NASA as candidates for Phase B preliminary design and definition studies, but only three were selected for detailed analysis. These evolved through continued analysis and by compromise into the Revised Baseline station NASA decided to build.

Two exceptional configurations among the five examined were eliminated by 1984. One was the Spinning Solar Array with gyroscopic stabilization.

TABLE 2.1. Cost Estimates for Block I Station
(in billions of dollars)

	Initial Estimate (1984 $)	Revised Estimate (1984 $)	(1988 $)
Research & Development	$12.2	$12.2	$14.6
Flight Telerobotic Serv.	0.3	0.3	0.4
Rescue Vehicle	1.5	1.5	1.8
Test Enhancement		0.0 to 2.5	0.0 to 3.0
Backup Hardware	_____	0.2 to 1.4	0.2 to 1.7
Total R & D	$14.0	$14.2–17.9	$17.0–21.5

The pressurized laboratory and habitat modules would be suspended in the center of the structure in which the spin would generate a measure of artificial gravity. Centripetal acceleration would provide structural stiffness to the solar power arrays. NASA engineers perceived several problems with this structure. It would be virtually impossible to assemble it in space and it could not be launched fully assembled by the shuttle. Crew members would experience different forces of artificial gravity in different parts of the station, and the need to adapt to these forces presented a health concern. Moreover, the design would not provide the requisite microgravity environment for industrial and scientific experiments. Finally, this configuration could not be developed further by adding more power and pressurized volume.

The second configuration eliminated from early consideration was called the Big T. It had a large solar power array from which depended station elements. It was ruled out because the solar power array had only one degree of movement to track seasonal variations of the sun. Also, its development potential was limited.

The three designs considered for Phase B definition were the Delta, Planar, and Power Tower. The Delta had a large, sun-fixed, rotating structure in the shape of a delta consisting of a tetrahedron truss. One side of the structure was covered by the solar power array. The habitable modules were attached to the opposite side of the delta. The solar array would rotate to face the sun to draw power as the station circled the Earth. The Delta was structurally stiff, had a large enclosed service area and another large area for attaching scientific payloads. These plus factors were counterbalanced by reduced Earth viewing at points in the orbit and the need to provide a number of remote manipulator systems for servicing because it was not feasible to provide a track on which the manipulator could move. Simultaneous Earth and celestial viewing would be difficult with this design.

The Planar design, a presentation by the Planar-Concept Development Group, offered an inertially stabilized modular configuration in which pressurized modules would be attached to a transverse structural boom. Power would be provided by four gimbaled solar arrays. This design offered versatile viewing. Because of the inertial flight mode, the solar arrays could not be pointed continuously toward the sun.

The Power Tower presented the fewest operational constraints of the three designs NASA considered for Phase B. It had a single vertical keel with solar arrays mounted at one end and the modules attached to a boom at the "bottom" end, closest to Earth. This configuration provided gravity gradient stabilization, a technique first studied in orbit during the Gemini program in the early 1960s. The term *gravity gradient* refers to the manner in which the force of gravity varies on space vehicle flying perpendicularly to the surface of the Earth, so that one end is closer to the center of gravity of the Earth than the other. Because gravitational force is greater at the lower end, it tends to stabilize the vehicle and damp out yaw motion.

All three of these configurations studied in 1983–1984 could be assembled in Earth orbit, and the costs of all three were similar. The dual keel evolved so quickly from the Power Tower that both configurations were presented publicly at about the same time. The one great advantage of the Dual Keel was its greater torsional stiffness and capability to be adapted as a base for lunar and planetary flights requiring space ship assembly in orbit. This feature was sacrificed in the compromise that led to the Revised Baseline, but it may be restored in Phase 2 of the station.

T H R E E

THE GRAND DESIGN

O**N** December 1, 1987, NASA Administrator James C. Fletcher announced that four aerospace firms and their contractor teams had been selected to design and build America's international space station. Participating in it would be the 11 nations of Western Europe comprising the European Space Agency, the Dominion of Canada, and Japan.

Politically, the station would serve the free world as a permanent orbital establishment in parallel with *Mir,* the orbital establishment of the Soviet Union and the communist world. It was NASA's intention to commence operating the station as a permanent manned facility in 1995. Thus, by the end of the twentieth century, each of the planet's two space-faring societies would have achieved the means to establish a human presence in low Earth orbit. It was a first step toward human expansion into the Solar System in the twenty-first century.

Although the financing of this enterprise by the United States and its partners had not been settled, the manufacturing task was precisely allocated in the opening weeks of 1988. It was divided into four work packages, each managed by a NASA center.

When the elements of the station were fabricated and tested, they would be carried up into orbit sequentially by the shuttle and assembled by astronauts working out of the shuttle orbiter cargo bay. The assembly process was scheduled to start in January 1995 and to be completed and ready for human occupation on a permanent basis by the end of 1996.

Each of the four work packages was divided into two phases to accommodate funding constraints. Phase 1 covered the construction of the modified station. It consisted of three laboratory modules and one habitat module mounted on a transverse boom—a single horizontal truss—262.5 feet long. As redesigned, the length of the main truss would reach 508.5 feet when power modules were attached at each end. The modified station would provide nearly as much research capacity as the dual-keel version. It would be equipped for microgravity industrial, pharmaceutical, and human physiological research and for observation of the Earth. In addition to the U.S. laboratory and habitat modules, it would carry European Space Agency and Japanese research laboratory modules and specialized Canadian servicing systems. The station complex would include NASA and ESA polar-orbiting observatories and experiment platforms. These would be linked electronically to/but would fly independently of the main station.

Except for the polar orbiting observatories, the

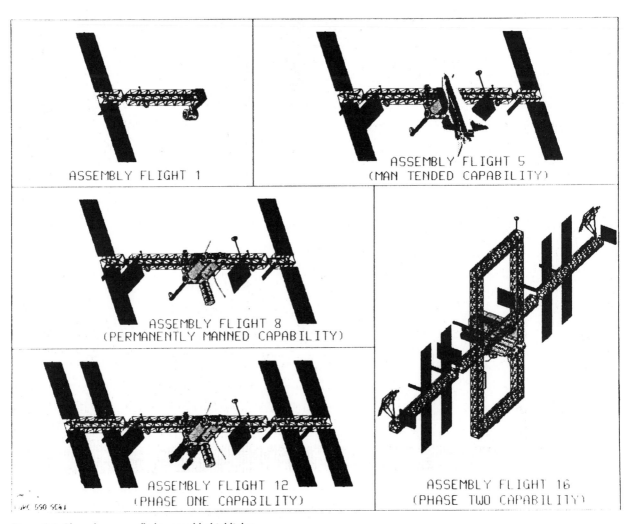

ASSEMBLY FLIGHT 1

ASSEMBLY FLIGHT 5
(MAN TENDED CAPABILITY)

ASSEMBLY FLIGHT 8
(PERMANENTLY MANNED CAPABILITY)

ASSEMBLY FLIGHT 12
(PHASE ONE CAPABILITY)

ASSEMBLY FLIGHT 16
(PHASE TWO CAPABILITY)

Figure 3.1. Phased program flight assembly highlights.

U.S. international station and auxiliary platforms would orbit the Earth at an inclination of 28.5 degrees to the equator at an altitude of about 260 miles. This inclination is the latitude of the Kennedy Space Center, from which the shuttle construction flights would be launched. The inclination is convenient for lunar and interplanetary flights launched from Earth orbit rather than from the ground. A space station in that orbit provides access to the Solar System.

The Soviet space station *Mir* is in an orbit inclined 56 degrees to the equator, the latitude of its launch site at Tyuratam in the Soviet Republic of Kazahkstan. It overflies northern and southern temperate zones while the U.S. station would pass over the tropical and subtropical latitudes. Mir orbits at an altitude of about 270 miles.

Phase 2 of the work packages covers future enhancement of the revised space station by the addition of upper and lower transverse booms and vertical dual keels to form the picture frame architecture of the earlier station design. The additional truss work would provide attachment points for external payloads, increased electrical power, a free-flying unmanned experiment platform co-orbiting with the station, and facilities for servicing an Orbital Maneuvering Vehicle and satellites.

Although NASA space station policy anticipates

Figure 3.2. This full scale model of the Freedom Space Station was erected at the Johnson Space Center, Houston as a demonstration. Shown at left rear is the Japanese Experiment Module and at right rear, the U.S. Habitat Module. The European Space Agency's Columbus Module is shown at left front and the U.S. Laboratory Module is shown at right front. The four main modules are interconnected by center nodes with airlock access. (NASA)

utilizing the station as a base for assembling manned and unmanned vehicles for lunar and interplanetary flights, the structures required for this purpose were not included in the Phase 2 enhancement. So far as the phased construction program was laid out in 1988, the functions of the station were scientifically, commercially, and industrially Earth oriented.

Although the space station is considered in NASA planning as a "node" for lunar and interplanetary missions, the space agency did not include the provision for this function in its request for proposals from industry. Subsequent designs do not show it. However, the node or space base function is not precluded from being added to the Phase 2 enhanced station, especially if an Orbital Transfer Vehicle (OTV) is based at the station as an upper stage. With the OTV, payloads could be moved from the station to geosynchronous orbit some 22,000 miles higher or to lunar orbit and beyond.

THE CONTRACTORS

The four principal contractors selected for Phase one of the work packages were the Boeing Aerospace Company of Huntsville, Alabama; the McDonnell Douglas Astronautics Company of Huntington Beach, California, and Houston, Texas; the General Electric Company of Valley Forge, Pennsylvania, and East Windsor, New Jersey; and the Rocketdyne Division, Rockwell International Company, Canoga Park, California.

The total cost projected by the four was $5 billion for Phase 1 and $1.5 billion for Phase 2. Phase 1 was considered a 10-year project from its start through a year after completion. Phase 2 was considered a priced option.

Work Package One, managed by the Marshall Space Flight Center, called for Boeing to build the U.S. laboratory and habitat modules, resource nodes by which the modules are connected, logistics ele-

Figure 3.3. In this cutaway view of the space station laboratory module, scientists are shown operating materials processing equipment. The 44½ foot long module, pressurized for shirt sleeve comfort, provides 6533 cubic feet of work space. The nodes interconnect the laboratory module with the habitat module beside it also provide habitable space for work and storage. (NASA-Boeing Aerospace)

ments, airlock systems, internal thermal controls, audio and video systems and associated software. Boeing cited costs of $750 million for Phase 1 and $25 million for Phase 2 options.

Work Package two was under the management of the Johnson Space Center. It called for Mc-Donnell Douglas to provide the truss structure to which the modules, solar power arrays, external experiment packages, instruments, and Mobile Servicing Transporter would be attached. The package included the communications and tracking system and the propulsion system for maintaining the station in orbit at required altitudes. Projected costs were $1.9 billion for Phase 1 and $140 million for Phase 2.

Work Package three was managed by the Goddard Space Flight Center at Greenbelt, Maryland. Its contractor was General Electric, which was required to develop the polar orbiting platform, provide payload attachment hardware, integrate the Flight Telerobotic Servicer with the structure as it was assembled, and develop a satellite-servicing system. The satellite-servicing system would not be implemented until Phase 2 development was approved. Projected costs were $800 million for Phase one and $570 million for Phase two.

Work Package four, managed by the Lewis Research Center, called for Rocketdyne to design and build electric power systems for the station and the platforms. The Phase 1 power system used photovoltaic cell arrays to generate 75 kilowatts of electric power from sunlight and batteries to store power to keep the station operating on the night side, or umbra, of the orbit. The projected cost of the system was $1.6 billion. The system would consist of two 123 foot long power modules attached to the ends of the transverse boom by "alpha" joints. The joints would rotate the modules to point two pairs of photovoltaic arrays toward the sun as the station moved around the day side of the planet. The solar arrays consisting of eight solar cell panels on each module, extend perpendicular to the boom like the wings of a gigantic moth, with a spread of 208 feet.

The Phase 2 work package would add a solar dynamic power system based on a sun-heated turbine generator using the Brayton cycle. This power

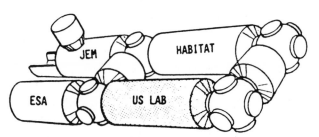

Figure 3.4. In this sketch, provided for Congress and news media, NASA listed the functions of the European Space Agency and Japanese modules as well as the U.S. habitat and laboratory modules. The roles of the Marshall, Johnson, Goddard and Lewis Centers of NASA in the development of the station and Canada's Mobile Servicing Center are depicted. (NASA)

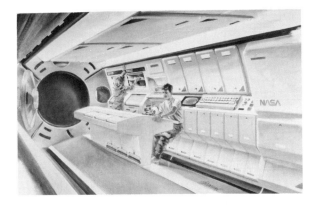

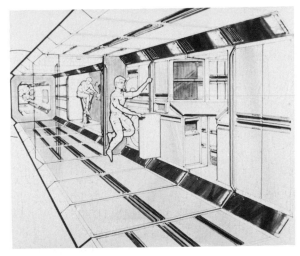

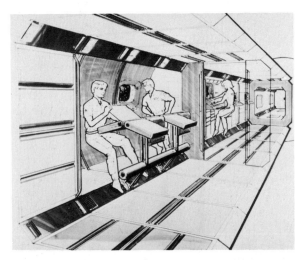

Figure 3.5. These sketches depict crew persons in the habitat and laboratory modules. Some areas can be closed off for privacy in the habitat where a Ward Room arrangement is suggeted (A). The habitat module housing a crew of eight provides at least as much space per person as a nuclear submarine. Tours of duty aboard the station have been tentatively set at 180 days. (NASA)

system would produce an additional 50 kilowatts at a cost of $740 million.

Each of the major contractors had enlisted sub-contractors. Boeing's team included Teledyne Brown Engineering, Huntsville; Lockheed Missiles and Space Company, Sunnyvale, California; Hamilton Standard, Windsor Locks, Connecticut; Garrett Airesearch, Torrance, California; Grumman Aerospace Corporation, Houston; ILC Space Systems, Houston; and Fairchild–Weston Systems, Inc., Syossett, New York.

The McDonnell Douglas team included IBM, Houston; Lockheed; RCA Corporation, Camden, New Jersey; Honeywell, Clearwater, Florida and Astro, Carpenteria, California. With General Electric in Work Package three was TRW, Redondo Beach, California, and with Rocketdyne in Work Package four were Ford Aerospace and Communications Corporation, Palo Alto, California; the Harris Corporation, Melbourne, Florida; the Garrett Corporation, Tempe, Arizona; and the General Dynamics Corporation, San Diego, California.

CONSTRUCTION IN FREE FALL

The space station is the first large structure humans have ever planned to build in orbital free fall. The unique aspect of this environment is the absence of terra firma. The station would be anchored to an orbital metal truss as long as a football field. It would be pieced together by astronauts standing in foot restraints and secured by tethers in the shuttle orbiter cargo bay or on its remote manipulator arm. Could it be done? Astronauts Sherwood Spring and Jerry Ross had shown how in the EASE and ACCESS experiments aboard the shuttle orbiter *Atlantis* in the fall of 1985 on Mission 61-B. Whether the task could be made simpler and accelerated by the Flight Telerobotic Servicer remained to be seen. The servicer itself could introduce complexity into the construction process.

This method of piecing together a large structure in orbit seemed to suggest building a bridge in heaven with an erector set. It had been ordained by NASA's decision in 1970 to abandon the expandable Saturn 5–Apollo space transportation system and substitute the reusable shuttle for it. As a result, the United States traded a reliable, heavy lift launch vehicle for one that failed on its twenty-fifth launch and crashed in the sea, killing the crew. Still, the shuttle was remarkably flexible. The defective solid rocket booster sealing system was reengineered, the main engines were improved, landing gear and brakes were made less likely to fail, and Congress appropriated funds to replace *Challenger* with a new orbiter. By 1992 the fleet would be whole again. But America lacked a heavy launch vehicle. It would take more than a year of shuttle flights to put into orbit the tonnage the Saturn 5 could lift with a single launch.

THE FLIGHT CREW'S TASK

In its request for proposals (RFP) to contractors, NASA headquarters characterized the method of building the space station in orbit "as a significant advance in the methodology of building a manned spacecraft." In sum, that was what the space station was—a manned spacecraft. The one NASA had in mind would orbit the Earth, but its capabil-

Figure 3.6. A full scale mock-up of the space station common module or habitat was built in 1985 by Martin Marietta Aerospace for design studies. The module is 37¼ feet long and 14½ feet in diameter, slightly shorter than the recent version. The astronaut mannequin suspended above is riding the MMU. The wooden ramp at the rear confirmed the atmosphere of an exhibit. Note the nautical style porthole forward.

ity went beyond low Earth orbit. With a propulsion system, it could fly to the Moon or Mars. The RFP noted that:

"The process of assembling, maintaining and modifying a facility while it remains in low Earth orbit and continues to be operational will require changes to the traditional method of building spacecraft . . . it is not possible, for example, to bring the space station back to the launch site for retrofit after a single mission or after several years of operation; neither is it possible for scores of technicians to be made available for routine maintenance. Once the launch process begins, all operations must be accomplished within the resources provided by the flight crew."

NASA warned the contractors: "Recognition of these functional changes must be made throughout the design, development, manufacture, test and certification phases and new processes must be introduced which will assure that operations, maintenance and field modifications can be accomplished on orbit."

In terms of construction resources, NASA recognized that it was handicapped by the loss of *Challenger* but insisted that assembly of the station "can be successfully deployed with the current system."[1] The current system consisted of three orbiters capable of flight: *Columbia, Discovery,* and *Atlantis. Enterprise,* designed for approach and

Figure 3.7. An alternate power concept is shown here by an artist for TRW, Inc. It depicts a solar array concentration system with deployable thermal radiators as part of the station's utility module. TRW, Inc.

Figure 3.8. Power for the station would be provided by this Solar Thermal Dynamic Power generating system designed by the Harris Corporation, Melbourne, Florida. Sunlight is concentrated by an array of hexagonal mirrors configured into a huge parasol 64 feet in diameter and focused on a working fluid that drives a generator. A pointing control aims the system at the sun. (Harris Corp.)

landing tests in 1977, was not equipped for orbital flight.

The space station construction program projected 18 shuttle launches, 12 for assembly of the main base and six for logistics and outfitting. All of them would be made from two pads at the Kennedy Space Center. Vandenberg Air Force Base in California had been considered earlier in the 1980s as a shuttle launch site for polar orbiting missions, but the space station in a subtropical obit could not be reached from there. Expendable rockets would launch the U.S. polar orbiting platform from Vandenberg. The ESA polar orbiting platform would be launched from Kourou, French Guiana, by Ariane, Europe's rapidly developing middle-weight launcher.

To ease pressure on the shuttle fleet, some logistics and outfitting payloads for the space station could be lifted by expendable rockets from Cape Canaveral. But even with this option, the freight schedule would keep the shuttle fleet busy. Four launches were scheduled in 1994, six in 1995, and eight in 1996 for the space station. This launch rate does not appear excessive, however, compared with pre-*Challenger* accident launch rates. Starting with two launches of *Columbia* in 1981, the launch rate rose to three in 1982, four in 1983, five in 1984, and nine in 1985. The nine missions in 1985 were flown by three orbiters—*Challenger, Discovery,* and *Atlantis*—while *Columbia* was being refurbished. A tenth flight was in prospect in 1985 with the return of *Columbia* to service, but weather and minor propellant problems delayed it until January 12, 1986. The *Challenger* launch on January 28 was the second for 1986.

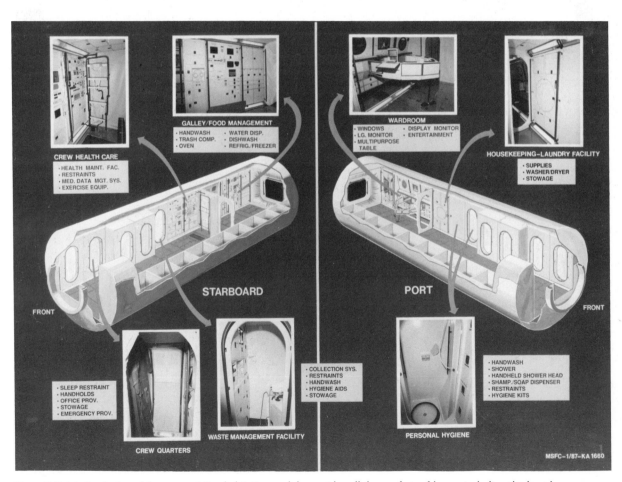

Figure 3.9. Interior design of the space station habitation module provides all the comforts of home, including the laundry.

In view of flight rate experience, the launch rate for space station construction appears conservative and would readily allow scientific and commercial missions in addition with a four-orbiter fleet. It is probable that the new fourth orbiter (OV-5) will be on line during the station construction period. The traffic model projected in 1988 called for one flight in that year, 7 in 1989, 10 in 1990, 10 in 1991, 12 in 1992, 13 in 1993, and 14 in 1994 (fiscal years).

In April 1988, NASA headquarters announced plans to develop an advanced solid rocket booster that would enable the shuttle to lift more tonnage to low Earth orbit and an extended-duration orbiter that would double orbiter flight time.

THE MODULES

Work Package one under Marshall supervision covered the two principal American modules, the laboratory and the habitat. A crew would work in one and eat, relax, and sleep in the other. The modules, aluminum cylinders about 44 feet long and 14 feet in diameter, would provide a shirt sleeve environment through an environmental control and life support system. Water and atmosphere would be continuously purified and recycled. Ultimately, such a system would become the prototype of life support technology for manned interplanetary flights and for expeditions landing on the Moon and Mars.

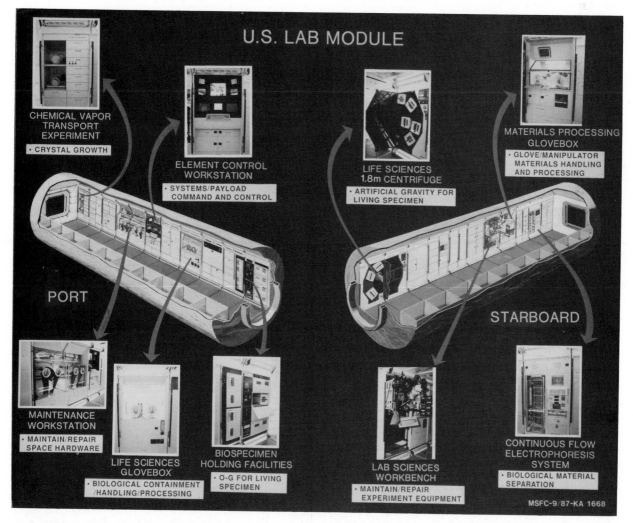

Figure 3.10. Interior design of the U.S. laboratory modules continues experiments in materials processing and pharmaceuticals previously carried on the shuttle. An important post-shuttle feature is a centrifuge for experiments in artificial gravity, a possible requirement for long, interplanetary flights. (NASA Marshall Space Flight Center)

A sophisticated life support technology based on recycling major elements of the biosphere begins with the space station.

The modules may be compared to a deluxe mobile home. NASA designers maintained that the habitat module can comfortably accommodate a crew of eight men and women. The module is being designed to have facilities for eating, sleeping, personal hygiene, waste management, recreation, and physical fitness. The astronauts would be able to monitor their health with x-ray machines and equipment to analyze blood and urine.

The habitat module galley would provide a stove, refrigerator, and food storage pantry. It would have facilities for washing hands and compacting trash and a dishwasher. The bathroom would provide for human waste collection, a shower, a sink, for washing hands and face and a drying facility. The habitat would have a wardroom with a multipurpose table and a laundry. Crew quarters were designed with the experiences of the shuttle orbiter and *Skylab* as guidelines. Each crew member would have a separate compartment, with a curtain for privacy and storage space for clothing and other personal articles.

The ESA module, *Columbus*, will be 41 ft long

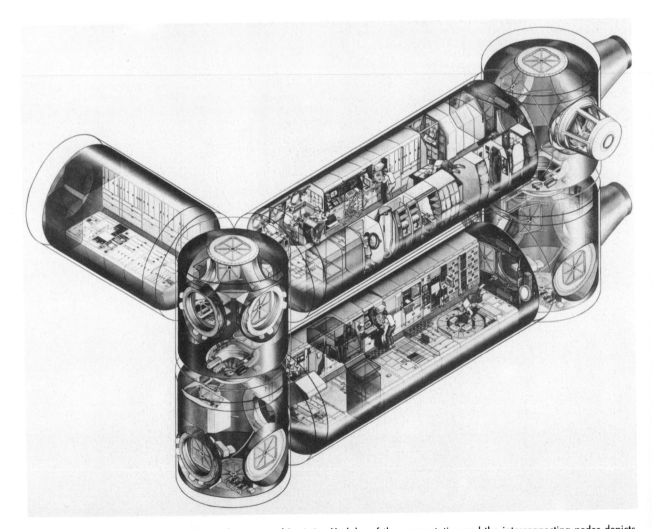

Figure 3.11. This cutaway of the Habitat, Laboratory and Logistics Modules of the space station and the interconnecting nodes depicts the "up and down" design of the module interiors. Up and Down orientation is carried out in the entire station for crew comfort and would be reinforced in the station by views of the Earth. (Boeing Aerospace).

and 14 ft in diameter. Its four spacelab segments are elements of an all-welded primary structure with docking ports at each end and a scientific airlock to expose experiments to vacuum and transfer tools and equipment.

The Japanese Experiment Module is 35 feet long and 14 feet in diameter. It consists of a permanently attached laboratory, an open platform, and a Logistics Module. The Experiment Module and Logistics Module are pressurized.

Station design thus calls for a cluster of four modules—two American, one European, and one Japanese—centered on the transverse truss. They are connected by structures called *nodes*. These are pressurized cylinders 14 ft in diameter and 17 feet long. They will provide additional pressurized space for storage, distribution systems, and berthing equipment for the shuttle or Logistics Module, and they will have attachments connecting them to the truss and the modules. Cupolas may be added to the node ports to allow a view outside. The nodes also will have docking mechanisms and hatches.

Node 1 between *Columbus* and the U.S. laboratory modules will be a control center for manned and unmanned operations. It will be attached to the hyperbaric airlock, a variable-pressure lock, and to Node 2. Both nodes will contain a berthing mechanism to permit temporary linkage to the Logistics Module. Node 1 may also contain elements of the propulsion system.

Node 2 will be a command and control station for man-tended operations. It will contain the airlock control station as well as berthing elements for the airlock. It is to be located between the Japanese Experiment and the U.S. Habitation Modules.

Node 3 will be the primary command and control center for the pressurized areas of the station. It will be placed at the forward end of the U.S. Laboratory Module and will contain berthing mechanisms for the shuttle orbiter. It may also have a back-up control station for the Mobile Servicer.

Node 4 may be the operations center and control station for the Mobile Servicer. It is to be attached to the forward end of the U.S. Habitation Module and connected to Node 3. It will provide pressur-

Figure 3.12. In the center of the laboratory module the station crew will find a general purpose workbench, with all the familiar clutter of the shop. (Marshall Space Flight Center. NASA).

ized passage to and from the modules and contain the mechanisms for orbiter berthing.

Pressurized and unpressurized logistics carriers will be used to transport consumables, equipment, and fluids to the station and return waste products, equipment, and experiment results in the shuttle cargo bay. The pressurized carrier will be 14 feet in diameter, but its length had not been determined in 1989. It would be berthed at the station at Node 1 or Node 2. The unpressurized carrier will transport propellant and equipment. It also will be berthed at a station port. Its design was not fully determined in 1989.

The Schedule. As projected by NASA's request for proposals, the station construction flight schedule called for launching the U.S. Laboratory Module January 1, 1995; the U.S. Habitat Module, May 1, 1995; the Japanese Experiment Module and Exposed Facility (first element) February 15, 1996; the ESA *Columbus* Module May 15, 1996; and the Japanese Logistics Module and Exposed Facility (second element) August 15, 1996.

NASA and ESA polar orbiting platforms would be launched by expendable rockets November 1, 1994 (from Vandenberg) and between May and July 1995 from Kourou. The NASA polar platform is designed to make continuous observations of the continents and oceans and of the sun and to monitor changes in the atmosphere. It would carry sensors to detect radioactive and nonradioactive ores. The ESA platform has been described as a general-purpose Earth observatory.

Both platforms essentially are satellites equipped

Figure 3.13. In this view of the laboratory module, the general purpose workbench is out of sight, in a work station. At the forward end is the centrifuge and storage racks for plants and animals. Materials processing equipment is at the aft end. (Marshall. NASA).

to carry on scientific investigations supplementing those conducted on the station. In the Phase program, starting in 1997 if funded by Congress, a second NASA and a second ESA platform would be added, this time in the station's low-inclination orbit, for astrophysics experiments and for Earth observation. If the Phase 2 enhancement should proceed, it would add the upper and lower booms, dual vertical keels, solar dynamic power, and extended logistics and servicing units in 1997–1998. For late 1996, ESA has plans to put a man-tended free flyer into orbit. It would be launched from Kourou by Ariane. It may be supported by the station.

Two principal construction devices are expected to have an important part in the assembly of the station. These are the Mobile Servicing Center, which consists of Canada's Mobile Remote Servicer and NASA's Mobile Transporter. The center is composed of a base mounted on the Mobile Transporter, a Remote Manipulator System "arm" like the arm on the orbiter, and a special-purpose manipulator for changing out orbital replacement units and attached payloads.

The Mobile Transporter slides along the truss and positions the Remote Servicer to perform its chores. The manipulator devices on the servicer can extract cargo from the orbiter payload bay and move it to the station assembly site. It can provide foot restraints and other positioning devices to give crew people a footing in orbital free fall. These devices provide the means to perform assembly work and inspect it while falling around the Earth.

This Mobile Transporter and Remote Servicer system can also move elements of station hardware to the work sites. The remote manipulator arms provide the means of handling and positioning hardware. The arms have "hands" (end effectors) that can perform the grasping functions of human hands. There is also a robotic special-purpose manipulator capable of changing out station hardware, moving attached payloads, and performing specific maintenance and servicing.

In preparation for assembling the station, NASA has been developing the Flight Telerobotic Servicer (as urged by members of Congress) to substitute for astronauts in doing maintenance and inspections of the station from the orbiter. This

Figure 3.14. The laboratory segment shown here contains plant storage racks and an animal habitat for microgravity studies of plant growth and animal behavior. (Marshall. NASA).

machine is designed to help with the assembly of primary structural elements by moving them into their fittings; erecting the boom structure; attaching utility trays; and installing fiber optic lines, data management system lines, copper power cables, gaseous hydrogen and oxygen tubing, meteor shielding, thermal coolant lines, and radiators on the transverse boom.

The Telerobotic Servicer is being designed also to install photovoltaic (solar) cell arrays at the ends of the transverse boom and communications and tracking as well as guidance and navigation and control equipment in the primary structure. This machine also will mate two halves of the transverse boom and connect power, thermal, and reaction

control systems and data umbilical lines. It will mate the airlock to mechanical and utility interfaces on the nodes.

That is not all. The Flight Telerobotic Servicer will be able to change out a module in a free-flying spacecraft on orbit while the servicer is attached to a holding fixture in the shuttle cargo bay. It will be able to perform inspections from the shuttle's remote manipulator arm or the Space Station Mobile Servicing Center. The servicer is operated, of course, by astronauts using direct manipulative control or programming a sequence of commands from several work stations as the station assembly progresses.

Canada's Mobile Servicing Center and NASA's

Figure 3.15. In orbital free fall aboard the space station, crew persons can sleep standing up, relative to the up-down orientation of the habitat, as well as lying down. That is demonstrated by a technician in a sleep station. At left, another technician works at a terminal in an adjacent cubicle. (NASA).

Flight Telerobotic Servicer are the vanguard of a squad of robotic devices that may eventually build large and complex space structures with only nominal human supervision. The space station assembly process may demonstrate the potential of this technology.

The station will have semiautomatic devices. The Attached Payload Accommodations Equipment is being designed to mount and run external scientific equipment. It will provide instruments with power, fluids, and data links. The equipment could be used at any of the station's four utility ports or attachment points on the transverse boom. A precision pointing mount is to be provided for instruments, such as telescopes or sensors, that require precise pointing or continuous orientation in a particular direction.

The space station propulsion system is designed only to maintain orbit. The system is to carry four propulsion modules, tanks and fuel distribution systems. Each module will contain fuel tanks, plumbing and valving, a fuel pump, and two types of jet actuators. Hot gas actuators will provide thrust for orbit maintenance in the 25-40 pound thrust range and will use hydrogen–oxygen propellant. Smaller engines will be used for vernier stabilization in the one pound thrust range. A three-axis, low-thrust system will back up the Control Momentum Gyroscopes, the primary actuators of the station's Stabilization and Control System.

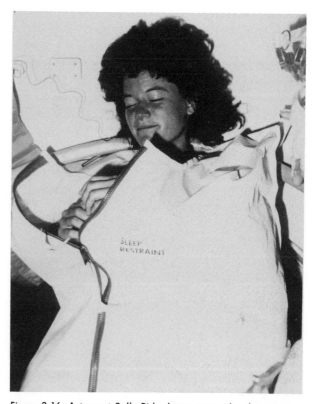

Figure 3.16. Astronaut Sally Ride demonstrates the sleep accommodations in the shuttle orbiter, in a sleep restraint device. She was captured by a fellow crew member at her sleep station in the Challenger mid-deck aboard the STS-7 mission in June 3 1983. Sleep stations aboard the space station are expected to offer more privacy. (NASA).

UPGRADING THE SHUTTLE FOR STATION DUTY

In the interim between the resumption of shuttle flights in the fall of 1988 and the start of space station construction in 1994 or 1995, NASA has plans to improve shuttle performance. Congressional committees in the House and Senate directed the space agency in 1987 to request financing for orbiter modification that would increase its flight time. An extended-duration orbiter program was recommended in 1987 by the National Research Council.

While an Extended Duration Orbiter would benefit space science, technology, and applications re-

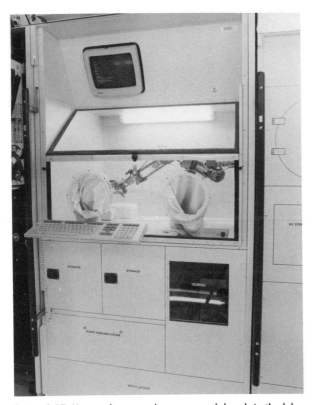

Figure 3.17. Next to the general purpose work bench in the laboratory is a planned glove box work station. It would enable researchers to handle materials with protective gloves or by a mechanical arm and hand. Glove boxes were used by researchers handling lunar soil and rocks at the Lunar Receiving Laboratory, Houston during the Apollo lunar exploration program. Contamination was feared until the Moon was found to be sterile. (Marshall. NASA)

search, NASA conceded, there was no urgency for it. The improvement would not make a direct contribution to the assembly of the space station.* It would, however, provide long-term support for each assembly mission. Still, it appeared that NASA headquarters regarded the project with several caveats, the agency's report showed.

The agency analyzed the cost and time it would take to increase orbiter flight time from seven to 16 days. The report concluded that the cost of developing one extended duration orbiter would be $126 million and would take 45 months. The report considered the participation of commercial firms in the program and in hardware costs but advised against it.

The essential modification required to increase flight time was to provide more power. The orbiter power system consists of three fuel cell power plants in which hydrogen and oxygen combine to produce electricity, with water as the by-product. The gases are stored cryogenically in a liquid state in three tans sets which provide enough fuel for four days. A fourth tank set was added to extend flight time to seven days. On the ninth shuttle mission, November 28–December 8, 1983, *Columbia* was equipped with a fifth tank set to extend flight to 10 days for *Spacelab*, which was carried in the cargo bay.

The report said that a 16-day mission would reduce the orbiter's designed payload capacity of 65,000 pounds by 8122 pounds on ascent and 4875 pounds on descent. About 7 ft of cargo space would be taken up by the installation in the rear cargo bay of a pallet with four hydrogen and oxygen tank sets. These would be in addition to the four used on seven- to eight-day missions to maintain the electric power level at 18 kilowatts. In addition to the added hardware and reactants, more nitrogen would be needed to make up atmosphere loss, and the capacities of the carbon dioxide removal and waste management systems would have to be increased.

The science data return to the extended-mission orbiter was estimated at 2.2 times that of a seven-day flight. The NASA report said there was no

---

* Report on the Extended Duration Orbiter, November 1987.

doubt that the extended flight time would allow more ambitious scientific and processing experiments and spacelab missions. It was noted that a 16-day orbiter would provide a significant safety margin for space station construction inasmuch as it would be available in orbit in the event of a station emergency.

The Advanced Solid Rocket Motor. The next major improvement proposed for the shuttle was the Advanced Solid Rocket Motor. It would increase performance and reliability, headquarters said. In the light of the National Research Council's opinion that the fleet's ability to meet the demands of assembling the space station was marginal, the proposal to improve the motor was submitted to congressional appropriation committees as a priority.

Motor improvement had been considered before the Challenger accident when evidence of field joint seal erosion and leakage of hot gas had been noted as a persistent problem. For more than a year NASA and Thiokol engineers had discussed methods of correcting the problem, but it was not considered urgent enough to justify the expense and time of redesigning the sealing system. Instead, the NASA directorate, intent on maintaining the launch schedule to meet commercial and military demand, waived fail-safe rules to fly the vehicle, despite continuing reports that the O-ring seals were leaking hot gas. Consequently, it was only after the Presidential Commission on the Space Shuttle *Challenger* Accident had completed its investigation that NASA acted to seek an improved and safer motor design. The agency promulgated a Solid Rocket Motor Acquisition Strategy and Plan March 3, 1987, calling for the advanced motor and presented it to Congress April 19, 1988. The advanced motor would replace the partially redesigned Morton Thiokol motor by 1994, the year that NASA planned to begin assembling the space station. Until then the shuttle would be launched by the partially redesigned motor with improved field and nozzle joint seals when flight resumed in late summer 1988. The advanced motor was intended to increase the performance of the redesigned motor by 17 percent, adding 12,000 pounds to vehicle's payload

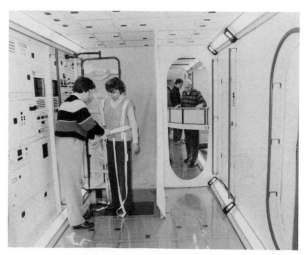

Figure 3.18. The space station will have a "sick bay" or health maintenance facility in the habitat module. Crew health will be monitored regularly. (Marshall-NASA).

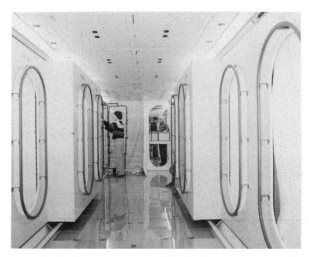

Figure 3.19. Sleeping quarters will be individual cubicles in the habitat module for eight crew persons. A shower and waste management system are provided in this section for the station habitat. (Marshall. NASA).

capacity on a launch to the shuttle construction orbit. As originally designed, the shuttle weight-lifting capability was 65,000 pounds to a 110 mile orbit, but that payload size was never flown because of landing weight constraints. Because of weight growth in the partially redesigned motor, payload capacity was reduced to 56,000 pounds. The advanced motor would be designed to lift 68,000 pounds to the station construction orbit. Inasmuch as landing weight would be substantially reduced by deliveries, it was not expected to be a constraint.

A monolithic booster design that eliminated the segments and the joints between them had been considered but was rejected by the design team. The hazard of filling the entire motor with 1,108,000 pounds of highly flammable propellant was considered unacceptable and unavoidable. Moreover, the Vehicle Assembly Building did not have a crane powerful enough to lift a monolithic booster rocket 149.6 feet long weighing 1,259,000 pounds. For these reasons, the advanced motor would retain the segmented structure that required the booster to be assembled, leak-checked, and inspected in the assembly building. There the boosters would be attached to the external tank on which the orbiter also would be mounted before the assembled shuttle was rolled out to the launch pad.

The design of the advanced motor eliminated factory joints, which would be replaced by welded joints. The change eliminated four Criticality 1 failure modes.* The field joints would still be made at the Kennedy Space Center but would have a new design. Instead of being joined by tang and clevis in the old way, the segments would be bolted together. This allowed the joints to be made, so that when the motor was fired, gas pressure forced the joints to close. The tang and clevis structure of the pre-accident joint allowed the joint to open under motor pressure, exposing the seals to hot gas erosion. The change eliminated 175 failure modes designated as Criticality 1R.†

The bolted structure would reduce dependence on the O-rings to stop gas leakage, the designers

---

*In the original design, factory joints were made at the Thiokol Plant and field joints at the Kennedy Space Center. Criticality 1 refers to a nonredundant failure mode that can destroy the vehicle.

†NASA briefing April 19, 1988. Criticality 1R refers to a redundant failure mode.

claimed. O-rings of rubber or of metal would continue to be used to seal the joints. The new joint would continue to allow for leak checking during assembly, as the old joint had. The assembly process would remain much the same as with the Morton Thiokol motor. Further, the advanced design would eliminate the requirement of throttling back the orbiter main engines as the shuttle passes through Max Q, the region where maximum dynamic pressure is exerted on the vehicle. Acceleration at this point would be modified by motor design. This feature would eliminate or reduce another 175 criticality 1 or 1R failure mode covering the orbiter main engines and Auxiliary Power Units, according to the designers. Also, redesign of the motor nozzle would reduce the number of nozzle parts and joints, thus removing another 16 Criticality 1 and 1R failure modes.

NASA presented the advanced motor design as a significant effort to enhance the safety and reliability of the shuttle by reducing any change of failure at launch, from which there was no escape for the crew. A pole which allowed astronauts to escape the orbiter by parachute in gliding flight was added to the crew cabin before the resumption of shuttle flights. The advanced motor was developed from concepts offered by five contractors: the Aerojet Solid Propulsion Company, the Atlantic Research Corporation, Hercules, Inc., United Technologies Corporation, and Morton Thiokol, Inc. The contractors were awarded contracts in August 1987 for Phase B definition and design studies and cost estimates. These formed the basis for the development program.

The NASA technical team concluded that the development of the new motor would require advanced production facilities which should be automated as much as possible. The team decided that modifying an existing plant would not be economical or productive. A new plant was required to meet the design objectives from the ground up. Preferably, said NASA, it should be located on a government-owned site, but operated by a private contractor. The space agency called for three types of proposals: the construction of a government-owned plant on a government-owned site, leased to a private contractor; erection of a privately financed plant

Figure 3.20. Although it looks like an experiment station, this part of the space station habitat module is the galley, according to the Marshall Space Flight Center. An oval table for dining or conferences will be included. (Marshall. NASA).

on a government site, to be operated by a private contractor; and erection of a contractor-owned plant on the contractor's site, provided the site was government approved and conformed to environmental impact restrictions.

The NASA Site Evaluation Board reviewed government-owned sites at the Kennedy Space Center in Florida, the John C. Stennis Space Center (formerly National Space Technology Laboratory) at Bay St. Louis, Mississippi, and NASA's Western Test Range at Vandenberg Air Force Base, California. Vandenberg was eliminated from consideration because political instability in Panama might interfere with the shipment of boosters produced in California through the Panama Canal. Shipment

across the country by rail was ruled out because of its accident potential. Another site in Mississippi, the Tennessee Valley Authority's Nuclear Processing Facility at Yellow Creek, was added to the list, on the assumption that since the site was closed it could readily be transferred to NASA.

Of the three sites, NASA clearly preferred the Kennedy Space Center, where thousands of acres of scrubland were available for a booster production facility. Such a site would establish booster production within a few miles or less of the Vehicle Assembly Building and effect savings in transportation costs and time. The time-saving advantage would prove critical if a space station emergency required an emergency launch of the shuttle.

Figure 3.21. A porthole is provided in the habitat ward room and a television screen showing what is going on at Mission Control, Houston . . . and the exact Greenwich Mean Time. (Marshall. NASA)

Whether government procured or privately built, the production plant would cost between $200 and $300 million, NASA estimated. The design, development, testing, and evaluation of the advanced motor would cost close to $1 billion, according to the agency's estimate. When production got under way, NASA analysts predicted that each flight set of motors would cost about $30 million, the current cost. Although automation would reduce labor-intensive production work, it might not reduce cost. Once the advanced motor production was established in a government plant, NASA would encourage the use of plant facilities by private enterprise beyond NASA's requirements. The production prospects of such a plant could be measured in terms of the 30-year life expectancy of the space station. Beyond that loomed proposals to develop solid-rocket-booster-derived heavy launch vehicles to match Soviet launcher development.

In June 1988 Morton Thiokol announced that it was dropping out of the bidding competition to build the advanced motor, and its executives said that the company would concentrate on continued improvement of the redesigned motor. It was possible that because of the cost of investing in a new plant, the advanced motor would be shelved and Thiokol would continue to build the solid motor booster for the shuttle indefinitely. It was also possible that the Thiokol motor would be improved to the point where it would equal the design performance of the advanced motor.

The Station Construction Schedule Slips. In April 1988, a congressional subcommittee warned there would be a reduction of about 50 percent in funds NASA requested for space station construction in the 1988–1989 fiscal years. It would force a delay in the start of station assembly by one year, from

Figure 3.22. A Martin Marietta Aerospace engineer inspects the interior of the company's full scale mock-up of the laboratory module. The mock-up built for design studies.

the first quarter of 1994 to the first quarter of 1995. The station assembly launch schedule consequently was revised and the permanent occupation of the station was set back at least a year.

This effect was announced by Administrator Fletcher in letters to Representative Edward P. Boland (D–Mass.), chairman of the House Subcommittee on HUD and Independent Agencies, and Senator William Proxmire (D–Wis.), chairman of the Senate Subcommittee on HUD and Independent Agencies. Fletcher's letter responded to the Joint Conference Report on Appropriations by the subcommittees. The report had ordered NASA to rescope (revise) and reschedule planned space station activities to meet the funding reductions. Fletcher balked at rescoping the station, which meant revising its configuration to cut costs. It had been cut back a few months earlier from the Dual Keel configuration to the Revised Baseline to reduce cost. Further changes would be "extremely unwise," Fletcher argued. The only reasonable way to meet the budget cuts was to slip the launch on the first station elements a year, he said.

Months earlier, Fletcher had notified the appropriation committees that severe cuts in station construction money would destroy the program. As Congress struggled with the fiscal 1989 budget in the toils of the nation's deficit, Fletcher conceded that with enough money to start construction, the space station, even though seriously underfunded, might be saved by stretching out construction.

A paramount objective of developing the space station as early as possible was to provide orbital research facilities for commercial users. That had been emphasized in President Reagan's Space Policy and Commercial Space Initiative promulgated February 11, 1988. Fletcher advised the appropriations committees that NASA would strive to ease the impact of a year's assembly delay by changing the assembly sequence of the baseline station. He said that the change would allow early man tending of the station to get experimentation started by occupying the laboratory module before the habitat module was attached. This arrangement required launch of the laboratory on the fourth shuttle assembly flight instead of the fifth, as originally scheduled.

With the fourth assembly flight, the station would be less than half built. It would consist of only one half of the horizontal truss, two of the four resources nodes, and use 18.75 kilowatts of electric power of the 75 kilowatts planned. The propulsion system would be available, however, to enable the partially built structure to maintain a safe altitude; otherwise, it would begin to descend, reenter the atmosphere, and burn. Provided also on orbit would be the docking system for the shuttle, the fluid management system, the communications and tracking system, and some of the laboratory equipment.

At this stage of assembly, man tending would be feasible; that is, astronauts could visit the station for short periods to start or monitor scientific or processing experiments. They could eat and sleep in the laboratory or in the shuttle orbiter until the habitat module arrived.

The space station directorate had opposed man tending as a phase in station development because it would delay completion and increase the cost of the permanently occupied station. The man-tending question had been raised by some House and Senate appropriations committee members in 1986. By 1988 NASA had changed its position. The agency conceded that an early man-tended capability could be provided by reordering the assembly sequence. Its new position was that man tending as a temporary expedient would not postpone permanent manned capability. Permanent occupation would be made possible with the arrival of the habitat module, Fletcher said, and a brief period of man tending would be feasible. In addition to the laboratory, the crew could dwell in a long-duration orbiter for a couple of weeks if necessary.

Fletcher said that the laboratory could be launched with "useful capability" but could not be fully outfitted until the sixth shuttle flight. That flight had been scheduled to deliver the partly outfitted habitat in May 1995. With the funding cut, the sixth flight would not be launched until early 1996 and would be limited to hauling laboratory equipment. The habitat module would be delivered later.

Fletcher held out hope that deliveries could be accelerated with the advanced solid rocket motor so that the laboratory could be fully equipped earlier, but in the dim light of the budget crunch prospects for the advanced motor and the long-duration orbiter did not appear bright.

NASA had projected the start of station assembly on the basis of the President's budget requests of $767 million in the 1988 fiscal year and $1.8 billion in 1989. However, because of the bipartisan budget compromise of 1988, NASA's budget allocation for the station was reduced to $425 million for fiscal 1988 and to $967 million for fiscal 1989. The underfunding in fiscal 1988 reflected the holdout of $121 million that had been appropriated for NASA in 1987 but had not been obligated. It was deemed available to NASA in 1988, but the congressional appropriations committees held it up by placing conditions on the release of $100 million. The main condition was that NASA had to agree to lease facilities in the proposed privately developed Industrial Space Facility.

NASA was saved from this dilemma by the timely intervention of its supporters in the Senate and House, who managed to have consideration of any commercial space development program postponed for a long-term study. The appropriations committees then released the 1987 funds to NASA except for $45 million.

The arbitrary allocations and reductions of space funds by the committees made the timely development of the station an uncertain prospect. In both houses of Congress, funds allocated to NASA in the administration budget came under attack by

Figure 3.23. A polar orbiting platform is to be flown in conjunction with the space station. This design is part of an $800 million work package assigned to the GE Astro-Space Division under the direction of the Goddard Space Flight Center. (GE)

House and Senate subcommittees on HUD and Independent Agencies. There was strong pressure to redistribute much of NASA's allocation to Housing and Urban Development to meet the need for low-budget housing in the Northeast.

During this period of budget turmoil in the spring of 1988, NASA's Associate Administrator for the space station, Andrew Stofan, resigned as of April 1, 1988. He had served less than two years, having been appointed in June 1986. Stofan had been director of the Lewis Research Center at Cleveland, an advanced technical research center. His departure seemed abrupt and he made no public explanation. This stirred speculation that he was fed up by efforts in Congress and the administration to push private research efforts in space at the expense of NASA and by funding oscillations and digressions that threatened to delay or destroy the space station. He was succeeded as station director by James Odom, director of Science and Engineering at the Marshall Space Flight Center.

THE NEW SPACE POLICY

Debate about federal subsidy for private space enterprise like the Industrial Space Facility had generated administration interest in the role of the private sector in space. At the outset of this development, Administrator Fletcher had opposed a NASA subsidy of the project as interference with the funding of the international station. As congressional support of a private station became manifest, Fletcher modified his stand and agreed to consider a plan whereby NASA might lease part of a private space facility.

At the outset of the 1989 fiscal year budget debate, NASA proposed legislation that would enable it to lease services on a Commercially Developed Space Facility. Under this arrangement the space agency would consider proposals from a number of companies that popped up demanding equal consideration with the Industrial Space Facility for a piece of the action in subsidized private space facilities.

Dissension over leasing and growing support for a study of its impact resulted in postponing action on this legislation, at least for the 1989 fiscal year. The threat of diverting international space station funding to subsidize a private station faded away. However, the role of private enterprise in space and the responsibility of the government to encourage its development remained unsettled.

In that regard, President Reagan moved to clarify the issue by promulgating a new national space policy and a commercial space initiative at the same time on February 11, 1988. The main elements of the policy were the establishment of a long range goal to expand the human presence and activity beyond Earth orbit into the Solar System; the creation of opportunities for U.S. commerce in space; and continuation of the national commitment to build the permanently manned space station.

A brief statement in the new policy calling for the expansion of the human presence in the Solar System was broadly interpreted in the spacework community as authorization to proceed with plans for the return to the Moon and a manned expedition to Mars. In order to lay a foundation for future exploration, the President requested $100 million in the 1989 budget for the development program called Project Pathfinder. It would provide the enabling technology for a range of missions, manned and unmanned, to the Moon and beyond.

Pathfinder would be focused on exploration and space operations technologies, the performance capabilities of humans in space, and the development of vehicles, such as the Orbital Transfer Vehicle and Orbital Maneuvering Vehicle. With the shuttle and the space station, these vehicles would provide a transportation infrastructure that made ambitious planning feasible.

Additionally, Pathfinder research would be directed toward the development of a closed-loop life support system, the technique and hardware required for aerobraking as a method of injecting interplanetary spacecraft into orbits around Earth and Mars, techniques of orbital transfer and maneuvering, the storage of cryogenic propellants (liquid hydrogen and liquid oxygen), and the assembly of large structures in orbit. The international space station would be the first large-scale construction in space.

The President said he would seek a three-year commitment from Congress for $6.1 billion to assure the sequential development of the station. In his policy statement the President emphasized the scientific potential and the applications of the station. The purposes of the station were broad and projected the expectations of two decades of planning.

They included scientific research in the microgravity environment, industrial and commercial processing experimentation on a broad scale, observation of the Earth and the Solar System, the assembly of vehicles in orbit for deep-space missions, the servicing of satellites, and the development of international cooperation and consultation in research among American, West European, and Japanese partners.

As part of his commercial space initiative, the President proposed the government's encouragement of private sector investment in the space station, calling on NASA to rely "to the greatest feasible extent" on private sector design, financing, construction, and operation of future space station development.

The policy called upon the government to facilitate private sector access to appropriate space-related hardware and facilities and to encourage the private sector to undertake commercial space ventures. It provided for the use of appropriate government facilities on a reimbursable basis and for the transfer of government-developed space technology to the private sector—consistent with national security.

The Department of Commerce was delegated to manage federal and civil terrestrial remote sensing operations (such as satellite imagery, environmental monitoring of lands and seas, and meteorological observation). The Department of Defense was charged with ensuring that the military space program would incorporate the support requirements of the Strategic Defense Initiative. Defense was to develop a "robust and comprehensive" antisatellite (ASAT) capability at the earliest possible date.

The policy set forth space transportation guide-lines based on a mix of the shuttle and expendable launch vehicles. The shuttle would continue to be operated and managed as an institutional arrangement, consistent with the current NASA/Department of Defense Memorandum of Understanding.*

Under the policy, NASA will provide launch services for commercial and foreign payloads only where the payloads must be man tended or require the unique capabilities of the shuttle or it is determined that launching payloads on the shuttle is important for national security or foreign policy. Commercial and foreign payloads will not be launched on government owned or operated expendable launch vehicles except for national security or foreign policy reasons. Civil government agencies will encourage a domestic commercial launch industry by contracting for necessary expendable launch vehicle services from the private sector or with the Department of Defense.

These policy guidelines were expected to have the effect of taking the shuttle out of the commercial satellite hauling business, except in special circumstances, and shifting that lucrative transportation to the private sector in the United States. Alternatively, domestic and foreign satellites could be launched commercially by other countries. During the 32-month hiatus in shuttle flights, commercial launch service was offered on ESA's Ariane rocket, China's Long March rocket, and the Soviet Union's Proton booster.

The new policy designated the Department of Transportation as the lead federal agency for developing federal policy and regulatory guidance for U.S. commercial launch activities. This would be done in consultation with the Departments of Defense, State, NASA, and other concerned agencies.

It was the government's intention, the president said, to lease space as an anchor tenant in a Commercial Space Facility, financed and constructed by the private sector. How the lease would be funded was not spelled out.†

---

*This arrangement provided that NASA would be responsible for operational control of the shuttle for civil missions and that the Department of Defense would control operations for national security missions.

†NASA had estimated that leasing 70 percent of an Industrial Space Facility for five years would cost the agency $700 million, *Aviation Week & Space Technology* reported January 11, 1968.

A Microgravity Research Board was established by presidential Executive Order to stimulate research into the commercial applications of the microgravity environment. The board was to include senior-level representatives from the Departments of Commerce, Transportation, Energy, and Defense; the National Institutes of Health; and the National Science Foundation.

The new policy also provided that the government would make available for five years the expended propellant tanks of the shuttle fleet at no cost to all American commercial and nonprofit endeavors for research, storage, and manufacturing in space. How these 155-foot-long tanks would be recovered and outfitted for any purpose was left to the imagination, but proponents of the idea believed recovery and reuse were quite feasible.

Conveniently, the tanks are dropped just before the shuttle reaches orbit velocity, so that they will not go into orbit but will fall into the Indian Ocean. To make them available for reuse, shuttles would have to retain the tanks during ascent until orbital velocity is reached and then release them. The tanks would have to be equipped with propulsion to maintain them in orbit while being outfitted as laboratories, storage facilities, or habitats. Unless some of this work could be done on the ground before the tanks were filled with propellant, converting them would require a shuttle mission. Several ingenious ways of refurbishing orbiting tanks have been worked out on paper napkins in space center cafeterias during the lunch hour.

The new policy proposed administrative steps to limit third-party liability in space activities, a concern of the commercial launch industry. The Reagan administration announced it would propose to Congress a cap of $200,000 on noneconomic damage awards to individual third parties resulting from commercial launch accidents. The liability of commercial launch operators for damage to government property would be limited to an insurance level required by the Department of Transportation. If losses exceed that level, the government will waive its right to recover the excess.

The administration expressed willingness in the policy to consult with the private sector on the potential of the construction of commercial launch range facilities separate from government facilities. This element of the policy opened the way for a private sector spaceport to be developed in east central Florida near the Kennedy Space Center and Cape Canaveral.

The overall objective of the President's Commercial Space Initiative was to stimulate the development of a civil space program by the private sector apart from NASA. The initial effort to attempt this by developers of a commercial space station encountered opposition because the project depended on a substantial handout from NASA. It thus threatened to compete for funds with the federal space station.

The paradigm of private sector enterprise in space has been the Communications Satellite Corporation, which evolved without government subsidy to develop a powerful, worldwide consortium. The Commercial Space Initiative set the table in space for private commercial sustenance in the twenty-first century, but there is another element of the new space policy that may determine its direction.

That is the call to extend the human presence and enterprise beyond low Earth orbit.

During the summer of 1989, rumors spread through the NASA-aerospace contractor community in Washington that several congressional committees were ready to cut $400 million from the space station budget for 1990–91. Administrator Richard H. Truly considered options to deal with the dilemma, if it materialized, by "rescoping" the station; that is, reducing its size and capability.

The rescoping would cut construction costs, reduce electric power generation, cut crew size from eight to four persons, and downgrade communications and tracking capabilities. These options, along with alternate station configurations that would save money, were assessed by an engineering team at the Langley Research Center.

The process was reminiscent of the "rescoping" of the space shuttle in the 1970s, which reduced its capabilities and its safety. Having been through that process, former NASA Administrator James Fletcher had warned Congress earlier that he would cancel the station program before he would compromise its efficacy by funding cuts.

Truly, however, found another way to deal with

the budget cut threat without rescoping the station configuration. He simply stretched out construction and assembly. However, the station assembly shuttle lift could start in 1995. It could, if necessary, be stretched out until 1998. This strategy still would achieve a fully operational space station with a full crew of eight by the end of the century.

Following a preliminary design review in 1990, the critical design review would be held circa 1992–93. Its outcome would determine when the contractors would begin manufacturing the modules and other station elements. Only time would be compromised.

On this basis, the long awaited space station could become a reality.

F O U R

T R A N S I T I O N

A long-term effect of the *Challenger* accident was a political demand for a reassessment of the nation's goals in space. It led to a realignment of national space policy two years later, when President Reagan extended the goal of human activities in space from low Earth orbit to the Moon and beyond. This reaffirmation of the Space Task Group report of 1969 endowed unofficial proposals for the return to the Moon and a manned expedition to Mars with the force of policy.

Having scrapped the Saturn 5–Apollo transportation system, which was capable of launching lunar and interplanetary flights, and replaced it by the shuttle, which was not, NASA faced the prospect of re-creating the infrastructure of lunar and interplanetary transportation in order to realize the new goal. The centerpiece of the new infrastructure was America's *fin de siècle* effort, the space station.

But the station that NASA was planning was not designed to function as a node or base for lunar and interplanetary missions. It was to be dedicated to microgravity research and development and to observation of the Earth. In this respect, station design expressed NASA's pre-accident outlook. The agency had no plans to send humans back to the Moon or anywhere else beyond low Earth orbit, and no capability to do so. Nevertheless, an articulate and restless group of scientists and engineers at NASA centers and federal and private research laboratories had persisted in developing conceptual programs for manned and unmanned explorations. The near focus was the Moon; the far one, Mars.

It was well known that the technological capability to implement these programs had existed for years in government laboratories and the aerospace industry, but the political will and a coherent plan were missing. In response to criticism that the national space program was characterized by ad hoc efforts and lacked long-range goals, President Reagan, with the support of Congress, appointed a National Commission on Space. Its mission was to propose civilian space goals for America that might be applied to define space activities for the twenty-first century. The commission chairman was Thomas O. Paine, former NASA administrator, who had directed the formulation in 1969 of the NASA document "America's Next Decades in Space." It had been the basis of a Space Task Group Report to President Nixon. But its proposals for space stations in Earth and lunar orbits, the development of a base on the Moon, and a manned expedition to Mars in the mid-1980s were ignored by the White House. Critics of NASA's fixation on low Earth orbit feared that recommendations of this scope by the National Commission might meet the same fate in 1986. But the explosion of Challenger and the loss of its crew of seven men and women had created a crisis of confidence in NASA. Not only did it damage the space agency's vaunted technical image, but the tragedy cast a pall over future space efforts. There was no alternative: NASA either had to go forward or collapse.

NASA was reorganized. Many of the Old Guard leadership at NASA headquarters and centers resigned or retired. James C. Fletcher, who served as NASA administrator in the Nixon and Ford administration from April 1971 to 1977, was recalled by President Reagan from an academic post at the University of Pittsburgh to head the space agency a second time.

Figure 4.1. Thomas O. Paine, former NASA Administrator (1968–70), National Commission on Space, 1985.

Figure 4.2. James C. Fletcher, NASA Administrator

In terms of policy, a significant feature of the agency's reorganization was Fletcher's addition of the Office of Exploration and an enhanced Division of Solar System Exploration. These developments responded to the ambitious report of the National Commission on Space, which called for a long-range program of establishing human settlements beyond Earth.

The commission's report, "Pioneering the Space Frontier," extended the American pioneering tradition and attitudes to the Solar System. It called for and described the technological infrastructure of a new age of the covered wagon that would carry a new generation of pioneers outward. Suggestions in Congress that Fletcher devise a NASA program implementing the proposals of "Pioneering the Space Frontier" put the new administrator on notice that the commission report would not be allowed to lie fallow. Fletcher responded to the report with one promulgated at headquarters. He appointed Astronaut Sally Ride, who had served on the Rogers Commission investigating the *Challenger* accident, to lead a planning group at headquarters with a mission of developing a long-range program for NASA.

Essentially, Ride's assignment was to distill from a collection of concepts and proposals in the files at headquarters and NASA centers a feasible projection of NASA activity. The orientation of the report was the regaining of NASA leadership in space. The long stand-down in shuttle flights during the redesign of the solid rocket motors, in contrast to Soviet advances in space station technology and operations, was telling the world that America had slipped—although the Russians had not yet landed men on the Moon, landed successfully on Mars, reconnoitered Jupiter, Saturn and Uranus, and flown a reusable shuttle. They were simply piling up more hours in low Earth orbit than Americans were able to do.

The Ride Report, entitled "America's Future in Space," was released by NASA August 17, 1987. It was done with a crispness, insight, and attention to detail that characterized the scientist–astronaut's performance as a crew member. The report listed four initiatives that NASA might undertake to assert leadership.

These were specific studies of the Earth from orbit (Mission to Earth); advanced reconnaissance of the Solar System by automated vehicles, including sample returns from planets, a comet, an asteroid, and the Saturnian moon Titan; the reestablishment of human presence on the Moon and construction of a base there and manned expeditions to Mars leading to future settlement.

With new emphasis on Earth observations for a space station, the Ride Report reiterated the main points of NASA's 1969 "America's Next Decades in Space" report to the Nixon Space Task Group. The 1969 report provided the practical guidelines for advanced lunar and planetary missions. It surfaced again in the post-*Challenger* period, with embellishment, as a means whereby NASA could regain its status and equilibrium.

The theme running through the Ride Report was the restoration of America's supremacy in space, a rationale reminiscent of the motivation for Apollo after the Gagarin and Titov flights. Beyond that, however, Ride described a detailed and practical program. In the proposal entitled "Mission to Planet Earth" Ride called for the establishment of a global system of observation from space. It would consist of free-flying platforms in polar, low-inclination, and geostationary orbits to observe long-term trends in the environment. The proposal called for nine platforms. Two American, one European and one Japanese would be in polar orbit. Three American, one European and one Japanese would be in geostationary, 22,300 miles high, hanging over the equator at various longitudes. Additional observatories would be established in low-inclination orbits (at 28.5 degrees). These would collect Earth radiation data. Surface images would be made by means of synthetic aperture radar from the shuttle.

The observatory system would measure global cloud cover; vegetation cover; ice cover; rainfall; ocean chlorophyll content and currents; motions of crustal plates comprising the Earth's surface; concentrations of carbon dioxide, methane, and other "greenhouse" gases in the atmosphere; and volcanic activity. The platforms and satellites comprising the observatory system would be designed and built to operate for decades and would be serviced by astronauts or by robotic devices.

Figure 4.3. Sally K. Ride, former astronaut, Presidential Commission on the Space Shuttle Challenger Accident member and Special Assistant to the Administrator, NASA.

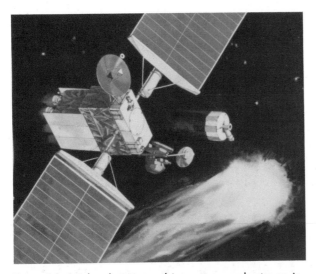

Figure 4.4. A solar electric propulsion system accelerates an instrumented probe toward a comet. Electric propulsion, the so-called "ion drive," expels ionized particles through a nozzle to provide thrust. The technology was developed at the Lewis Research Center, NASA, in the 1960s and tested successfully as a reaction control system on a satellite. The power source is a pair of solar cell arrays. (NASA-Lewis)

For the exploration of the Solar System, major missions would be planned for each of the planets beyond Earth, for their moons, and for comets and asteroids. The agenda included a comet rendezvous, an asteroid fly-by, a mission to Saturn, and a sample return from Mars. These missions would supplement flights already planned for the period 1989–1993: *Galileo* to Jupiter, *Magellan* to map Venus by radar, and a Mars-mapping observer satellite. The comet rendezvous and asteroid fly-by would penetrate the regions beyond Mars, in the main asteroid belt. The probe would be launched about 1993 on a cruise of six months. It would fly by the asteroid Hestia at 6200 miles and make visual and infrared light images of the surface.

The mission would continue on to the comet Tempel 2 and maneuver to within 15.53 miles of the nucleus. It would launch penetrators into the nucleus to obtain samples while flying along with the comet toward the sun. It would then maneuver away from the nucleus to observe the coma and tail of the comet.

The Saturn mission called Cassini, would be launched in 1998 and arrive at the ringed planet in 2005. An orbiter would make a three-year study of Saturn, its rings, moons, and magnetosphere; then three probes would be launched into the atmosphere. Another probe would be launched into the atmosphere of the big moon, Titan, which is suspected of having a nitrogen atmosphere denser than Earth's atmosphere. At Titan, the spacecraft would launch a semisoft lander to the surface.

At Mars, the sample return mission would be conducted on the surface by an automated roving vehicle similar to the Lunokhod vehicles the Soviets sent to the Moon. The rover would move along the surface and stop at selected places to collect soil samples. The samples would be launched back to Earth automatically.

Three Mars rover missions were contemplated, two launched in 1996 to ensure the return of a sample in 1999 and one launched in 1998–1999 to return in 2001. The samples would be delivered to low Earth orbit and picked up by an orbiting maneuvering vehicle that would deliver them to the space station for preliminary analysis. This process would avoid possible contamination of the Earth's

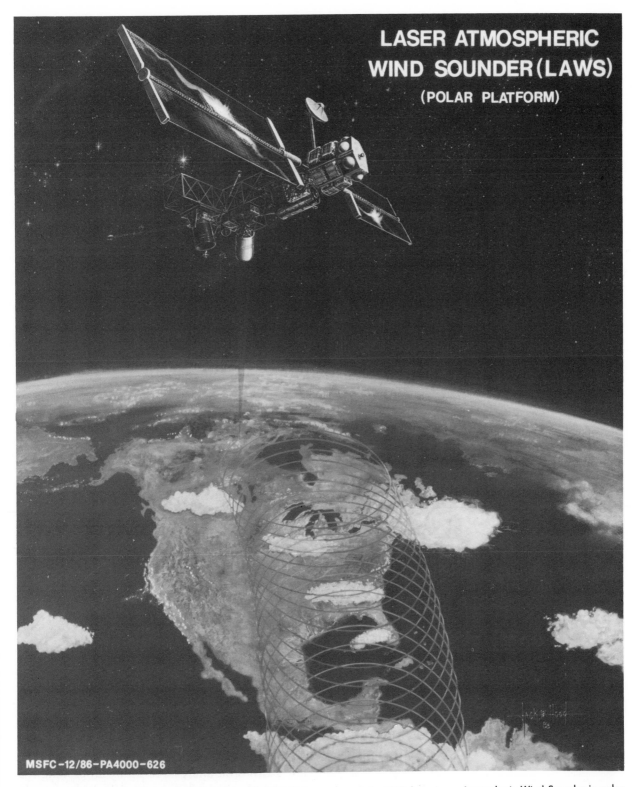

Figure 4.5. NASA's Marshall Space Flight Center called for design proposals in 1988 for a Laser Atmospheric Wind Sounder in polar orbit. It would measure the direction and speed of winds and density of aerosols by the reflection of light waves on particles in the atmosphere. (NASA-Marshall)

biosphere. The question of whether microbes exist on Mars was investigated by automated laboratories on two *Viking* landers in 1976, but the results of the tests were ambiguous and the question was not settled. In view of evidence for the former existence of free surface water on Mars, the possibility remains that microorganisms evolved there and may still exist.

A heavy-lift launch vehicle would be required for the Cassini mission to Saturn and also for a lunar initiative, the development of an outpost on the Moon. Astronauts could be landed during the year 2001. Their first step would be to learn how to extract oxygen from the rocks. When they accomplished that, they would have breathing gas and the major component of water and rocket engine propellant. Hydrogen would have to be imported unless equipment is brought to extract it by mining and heating from the lunar soil. Hydrogen is expected to be present in soil particles as fallout from the solar wind.

The first phase of the lunar base program would be the search for a site by the Lunar Geoscience Observer. The vehicle is to be equipped to map the surface, collect geochemical data, and look for ice at the lunar poles. Theory suggests that ice may have persisted in permanently shadowed polar areas since the formation of the Moon. Following up the Observer satellite, robot landers could make a detailed search of the polar regions and also collect

ADVANCED X-RAY ASTROPHYSICS FACILITY

MSFC–8/83–STE 3868B

Figure 4.6. A major new observatory planned for the 21st century is the Advanced X-Ray Astrophysical Observatory to probe x-ray sources in the universe. The project is managed by the Marshall Space Flight Center. (NASA-Marshall).

soil samples elsewhere for analysis. This phase of the lunar development program might be mounted in the early 1990s.

The next phase of the program, from 2000 to 2005, would bring astronauts to the lunar surface. They would travel from Earth to the space station by shuttle and then on to lunar orbit in a lunar transfer vehicle with equipment and supplies. They would descend from lunar orbit in Lunar Modules, as astronauts did in 1969–1972. They would remain on the surface up to two weeks to emplace scientific instruments, set up an oxygen plant, and devise a shelter. They would then return to Earth via the space station. By means of successive flights, the lunar outpost would be enlarged into a compound. It would include a habitat, research laboratory, soil-moving machinery, and oxygen extraction plants. The outpost could be developed within four or five years of intermittent occupation to support five persons for several weeks.

In the next phase, 2005–20, a sequence of five-year plans would enable the base to evolve into a permanently occupied site. By that time it would have a closed-loop life support system in which vital gases—oxygen, hydrogen, nitrogen, and water—would be recovered and reused. By 2010 as many as 30 persons would be living and working on the lunar surface for months at a time. The report forecast that lunar oxygen would be produced for propellant.

The transports required to establish such a base on the Moon included the heavy lift vehicle, to bring the main load of supplies and equipment from the Earth to the Moon; a shuttle fleet; a reusable Earth orbit to lunar orbit transfer vehicle; and a lander system.

AND THEN TO MARS

The Ride Report projected a Mars landing by the first American crew early in the twenty-first century, after orbital reconnaissance and sampling had been done. A landing site would be selected from data supplied by the Mars Observer satellite, added to data collected by the *Viking* orbiters in 1976–1977 and the *Mariner 9* orbiter.

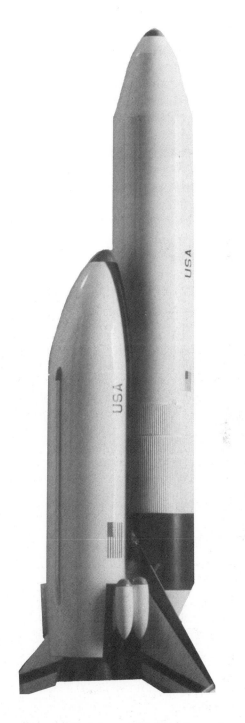

Figure 4.7. Since the abandonment of the Saturn 5, once the world's mightiest launch vehicle, the United States has lacked a heavy launcher for big payloads. One version of a heavy lift vehicle is depicted by this artist's concept of a launcher planned jointly by NASA and the Air Force. It would place 100,000 pounds in low Earth orbit and would be partly reusable. The first stage would be a fly-back booster and the second stage would be designed for recovery. (Boeing Aerospace).

The report made it clear that life science research on the space station was essential to validate the feasibility of long-duration human space flight in microgravity. The research would determine whether artificial gravity would be necessary to maintain crew health during the nine to 12 month trip. Three piloted round trips to Mars would be necessary before attempting to set up an outpost on Mars by 2010, the report suggested. The missions would be one-year "sprints," with crews spending two weeks on the surface.

It appeared that the Soviets were planning along the same lines. Their research into human adaptation to long-duration in microgravity was much further along than NASA, data—which were limited to the 84-cay *Skylab 4* mission in 1973–1974. On Christmas Day 1987, the Soviets announced that a crew had been flown to the *Mir* space station to start a year's stay in orbital microgravity. The tour would test countermeasures the Russians claimed to have developed that would enable cosmonauts on long space journeys to function when they reached their destination.

The Ride Report suggested that the piloted round trips to Mars could be divided into two parts. A cargo vessel loaded with fuel and supplies could be launched into a minimum-energy trajectory ahead of the crewed vehicle. While the cargo ship was en route, the crewed ship would be assembled in low Earth orbit and launched on a faster trajectory. It would carry a crew of six with support and propellant for the outbound journey.

In Mars orbit the crewed ship would dock with the cargo vessel, replenish its propellant tanks for the voyage back to Earth, and stand by in orbit while the crew descended in a lander to inspect the

Figure 4.8. This version of a heavy lift, advanced launcher with solid motors is depicted by General Dynamics. (General Dynamics).

surface. After a 10 to 20 day visit, the crew would return to their ship in the lander and go home, via the space station, where they would be required to spend time in a Rehabilitation Facility before returning to 1 g on Earth. The "sprint" mission would take about one year. Presumably, the cargo vessel would remain in Mars orbit. The report does not indicate whether it would be reusable.

The Mars mission would require a heavy lift vehicle to deposit fuel, equipment, and supplies in Earth orbit and the development of aerobraking technology as a means of reducing interplanetary flight velocity in order to fall into orbits around Mars outbound and Earth inbound. Aerobraking simply uses the atmosphere of a planet as a brake, instead of fuel. The aerobrake is a device that would be deployed on entry into the atmosphere to enhance drag and deflect frictional heat. Its use would be brief. When velocity is reduced to a level that would let the space ship fall into orbit around the planet, the ship skips out of the atmosphere and back into space. The operation is comparable to the flight of a flat stone skipping across the surface of a pond.

The report characterized the sprint missions scenarios as "a rational, sustained program leading to an outpost and eventual permanent base on Mars." These were steps in a "natural progression of human expansion" that leads "from the highlands of the Moon to the plains of Mars."

First, the Space Station. "America's Future in Space" projected a bold scenario for the twenty-first century, but its theme of "let's go forward" was muted by the 32 months grounding of the shuttle fleet and congressional dithering over space station funding. NASA's credibility as an effective research and development agency of government had fallen to the lowest level in its history. Five years after having declared the shuttle an operational transportation system, the redesigned boosters and the main propulsion engines were still undergoing development testing.

Because of the persistence of minor problems in the shuttle, the agency could not set up a coherent timetable for the resumption of flight and put out a

realistic manifest. Launch targets slipped from the first to the third quarter of 1988 for Mission 26, the flight of *Discovery,* as testing turned up problems in the redesigned boosters. Mission 26 was essential to put up NASA's second Tracking and Data Relay Satellite in geostationary orbit over the Pacific Ocean to replace the satellite lost with *Challenger. TDRS-1,* launched by an earlier *Challenger* flight in April 1983, was in geostationary orbit over the Atlantic. The two satellites were part of a system providing global communications and tracking for the shuttle and a number of satellites. The system replaced 14 ground stations in NASA's outmoded Space Tracking and Data Network.

In order to endow the programs projected by the National Commission on Space and Sally Ride's team at NASA headquarters with the prospect of probability, the space agency not only had to bootstrap its manned program back into orbit and build the space station, but also develop an advanced system of space transportation. It was a multibillion-dollar order. A new family of vehicles ranging from orbital maneuvering and orbital transport rockets to heavy lift boosters on the scale of the extinct Saturn 5 had to be created. The orbital maneuvering vehicle (OMV) had been funded for development in 1988, but the orbital transport and heavy lift vehicles had not then been authorized.

The launch pad of the post-*Challenger* thrust into the Solar System was the space station. Without it NASA lacked the means of finding out how astronauts would adapt to long-term exposure to microgravity. This effect could not be realistically simulated on Earth, even by long periods of bed rest. Consequently, effective countermeasures could not be developed without the permanently manned space station.

With a maximum of 84 days of data on continuous microgravity exposure by a crew in *Skylab,* it seemed unlikely that NASA could predict how a crew would fare on a nine-month voyage to Mars. The Soviet experience in the *Mir* station suggested that a cosmonaut could endure it, as indicated later in this chapter, but not how he would readapt to one-third Earth gravity on Mars. Without the space station and a variable-gravity research facility attached to it, adaptation to environments with less than 1 g would remain speculative until they were actually experienced.

Neither Phase 1 nor Phase 2 of the space station was planned to serve as a "node" or base for interplanetary or lunar missions, or for the assembly of large vehicles.* Both the National Commission and the Ride reports proposed separate stations for research and for base operations. NASA planners had considered combining these functions in a single station. Phase 2, with is large, picture frame truss consisting of dual vertical keels and upper and lower booms, could be developed for both research and

*Phase I refers to the single keel Revised Baseline station. Phase II refers to the advanced dual keel station configuration.

base operations. NASA would not make a decision until the Phase 1 station was a going concern.

Curiously, the design of the Phase I station could also be applied to an interplanetary transport. It was built for long-term habitation and experimentation. With adequate consumables and propulsion the space station can go anywhere in space.

The Space Station Applied to the Moon. Space station technology can be applied to the construction of a base on the Moon, some studies suggest. The station, even in Phase 1, could be used as a staging platform for lunar bound transports and provide vehicle assembly, check-out, and fuel storage services.[1]

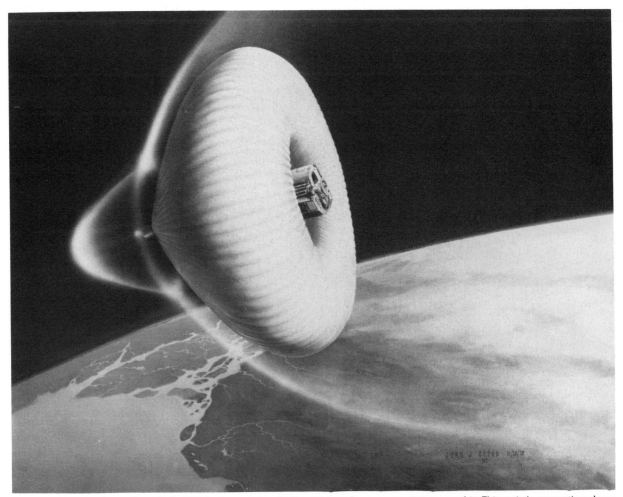

Figure 4.9. An orbital transfer vehicle (OTV) returns to low Earth orbit from lunar or geostationary orbit. This artist's conception shows how it is decelerated by a balloon-like parachute or "ballute" as it impacts the atmosphere. This method of airbraking enables the vehicle to rendezvous with the space station or a shuttle. (Boeing Aerospace)

Figure 4.10. One of NASA's major developments in the last two decades of the 20th century is the establishment of the Tracking and Data Relay System for space to ground communication. It supplants NASA's 25 year old network of ground stations with three communications satellites in geostationary orbit. These comsats will provide the major communications and data linkage between the shuttle, the space station and the ground, via a single station at White Sands, New Mexico. This photo shows one of the big satellites being inspected at the factory. (TRW)

One study considered a lunar base at a near-side equatorial location for a crew of three to five staying 30 to 90 days. The base would use lunar soil for radiation shielding but otherwise would depend entirely on resupply from Earth. The base structure would consist of space station habitat, laboratory and service modules, interface nodes, two roving vehicles, and a 100 kilowatt nuclear reactor for power. The interface elements would be derived from space station resource nodes. The base airlock would be similar to the space station airlock. A disposable logistics module would be used for resupply.

The lunar base would also adapt to its use the space station data management system and software; the communications and tracking system, including internal audio and video; the extravehicular radio and data collection equipment; and some

Figure 4.11. Tracking and Data Relay Satellite 3 is deployed from the cargo bay of the Space Shuttle Discovery on Mission 26 September 29, 1988. The satellite in a protective cocoon is attached to an Inertial Upper Stage rocket that boosted it to geostationary orbit at 22,300 miles altitude over the Pacific Ocean. This satellite replaced TDRS 2 which was lost in the Challenger accident. (NASA-Kennedy)

power equipment. Space station guidance and navigation and control equipment would be used for the lunar lander and ascent stage.

To function as a base for multiple lunar missions, the space station would require the addition of service facilities for lunar support vehicles and a fuel depot. Orbital Transfer Vehicles would be the primary means of transportation between the station and lunar orbit. Orbital Maneuvering Vehicles would be essential for on-orbit transfer. Space station lunar support facilities would include a Lunar Vehicle Assembly Hangar and a Propellant Tank Farm.[2] A heavy-lift launch vehicle would be required to bring up propellant for orbital storage.

A POT OF GOLD

In the post-*Challenger* era, the mission of the "new NASA" was openly and politically to regain world leadership in space—if, in fact, NASA had really lost it. The Soviets led the United States and its allies in exploiting the scientific and technological potentials of low Earth orbit, but the Moon and the planets challenged the energies and resources of both space-faring societies. If there was to be a race for gold and glory, it would come in the twenty-first century. Economic stringencies made it more reasonable to consider cooperative ventures. Competition or cooperation might indeed depend on what was found on the Moon.

NASA's new agencies were future oriented. The Office of Space Station would be the focal point of NASA activities for the balance of the century. The Solar System Exploration Division had the responsibility for coordinating the reconnaissance programs: the Mars Observer, the Venus Radar Mapper (Magellan), the Galileo mission to Jupiter, and the Cassini mission to Saturn and Titan. It was headed by Geoffrey A. Briggs, a physicist from the Jet Propulsion Laboratory, NASA's long-time center for planetary missions. The third headquarters addition, the Office of Exploration, evolved from the work of Sally Ride. Its function was planning manned missions to the planets, their moons, the asteroids, and the comets. The office was headed by an engineer from the Johnson Space Center,

John Aaron, who had been deputy to Dr. Ride. When she resigned to do research in her field at Stanford University, Aaron stepped in to the new office with a small staff and an assignment the size of the Solar System.

It was easy to confuse Aaron's Office of Exploration with Briggs' Solar System Exploration Division. The difference was marked. The division was an operating entity with an existing and well-planned program. The Office of Exploration as of mid-1988 had yet to find its way—by developing a plan.

As Briggs explained in the fall of 1987: "The agency (NASA) is in the process right now of looking to the future and trying to understand where it's going. It's just recently established a new office under the direction of John Aaron, Sally Ride headed it until recently, the Office of Exploration. The administrator sees it as a means of having the agency examine its long term future. And that office, working with other offices like this one (the Solar System Exploration Division), will presumably establish some requirements which in turn will be looked at by the office which has the responsibility for launch vehicles. They will look at the need for heavy lift launch vehicles, for this office and John Aaron's office, looking toward both unmanned and manned missions.

"There is also the Space Station. Each of the offices establishes its own desires and they all get put into a pot for Admiral Truly's people to look at." Briggs said that Truly, the Associate administrator for Space Flight, would determine the need for launch vehicles. Probably, the Exploration Division and the Office of Exploration would be the main drivers for the production of heavy launch vehicles once their programs were developed, Briggs surmised.

The charter of the Solar System Exploration Division, Briggs said, was to learn as much as possible about the origin of the Solar System and about the origin and early history of the Earth by examining those aspects of the other planets and take the lead in this investigation. The division had not received a significant charter to find and exploit resources out there; that might be too far downstream, Briggs said.

Figure 4.12. This artist's conception shows the tracking and Data Relay Satellite deployed in orbit. It transmits and receives in the S, C and Ku bands with seven antennas. (TRW)

Helium 3. In the case of the Moon, evidence of a vast potential source of energy had been found, Briggs said. The energy source was the recent discovery of a rare isotope of helium, helium 3, in the lunar topsoil. It was a deposit by the solar wind that had accumulated for eons. If it could be extracted from the lunar regolith, it would be more valuable than gold.

"The idea," said Briggs, "is that in the fusion reactor of the future, an ideal fuel would be helium 3, rather than a deuterium–tritium mixture. On Earth, you have helium 4, the normal state of helium. The isotope helium 3 isn't found in nature very much. You can create it, manufacture it at some considerable expense, but on the Moon, it has been soaked up for 4.5 billion years from the solar wind, absorbed by the lunar surface. So if you would go to the Moon and heat up the material you'd bake out the gas, including helium 3, along with helium 4, hydrogen, nitrogen, all the gases. But the helium 3 is a very precious material. It doesn't produce radioactive products in fusion re-

actions. That gives you an advantage in terms of the waste you would have to dispose of. Also, the products of the fusion reaction are charged particles that can be converted into electricity very efficiently. This is something people continue to work on.

"The helium 3 concept could make going to the Moon a very interesting thing to proceed with."

Briggs foresaw an interactive relationship between his division and Aaron's Office of Exploration. Unmanned missions might lead you to certain kinds of manned missions, he said.

"Meanwhile, people continue to think of the Moon as a means of locating certain kinds of astronomical telescopes, radio [telescopes] and otherwise. Our thinking in that area will continue to evolve.

"One of the next missions we will try to get started around here will be a Lunar Orbiter to map the crust both for scientific and inventory (purposes). We will continue making surveys of the asteroids. They are a bit of a tricky problem, there are so many of them. We have established a policy

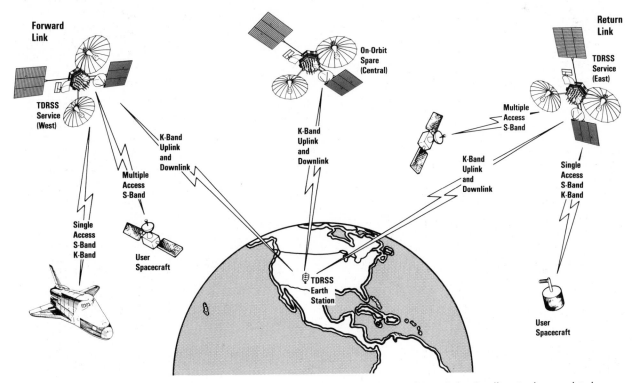

Figure 4.13. This diagram shows the general location of each of the three Tracking and Data Relay Satellites in the completed space system. The central satellite over the United States is a spare. Only the Atlantic and Pacific satellites would be functional in the system. (TRW).

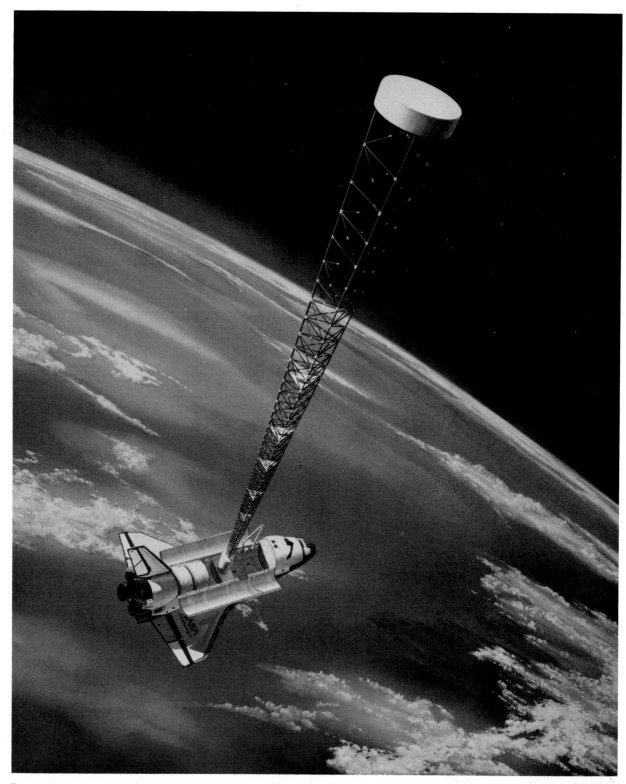

Figure 4.14. An experiment in the deployment of large structures in space has been designed by Harris Government Aerospace Systems Division for the space shuttle. Visualized in this artist's concept is a 200 foot deployable mast. The technology was first tested in 1985 aboard the shuttle orbiter Atlantis (see chapter II). (Harris, Melbourne)

that for want of a spacecraft to cross the asteroid belt, we will take on an asteroid encounter and give it proper priority. Our comet rendezvous mission has an asteroid fly-by in it. We may with one or two encounters per mission build up an inventory as we cross the asteroid belt.

Briggs emphasized the "writ" of his agency ran only to scientific exploration. It does not have a logistical responsibility to plan for manned exploration, nor for exploitation. Perhaps after 20 or 30 years of scientific investigation, sufficient information and understanding of what's out there would be accumulated to justify prospecting. "Presumably, one would hope to see a linkage between the unmanned program and the manned program of John Aaron," said Briggs. "My program office provides the direction to JPL. Much of the inspiration for new ideas comes from the centers. Headquarters provides people to take all of these ideas, distill them and then lay out a program."

The influence on these two headquarters exploration agendas of the programs proposed by the National Commission on Space did not seem strong. Briggs said that there was a need for NASA to respond "somewhat to the recommendation of the commission," but he added, "I'm not sure what the administrator is planning in that respect." Perhaps at some time in the future, NASA's response would be given to Congress, he said.

"There would not be any formal presentation to Congress by the administrator of NASA," Briggs added, "until NASA or the White House decided what they wanted to do."

Looking at the Mars Rover Sample Return Mission, Briggs observed that it represented a cost on the order of that for the new super collider (atom smasher). It would be hard to justify one versus the other, he said, but what was driving the Mars mission was not pure science, but competition. "What's driving things in this arena is the fact that we are not alone in this world. The Soviets have in fact charted Mars as the focus of their exploration activity.

Given the fact that people look to outer space endeavor as a reading of the vitality of a country, I think that the element of competition with the Soviet is one of the things that would be a factor in deciding whether the nation wanted to have another push at Mars. Because a mission like that is going to cost 5 or 10 times more than a typical planetary mission. National leadership in space becomes a political rather than a scientific issue. And people will often spend money for political reasons that they would never think of spending for other reasons."

THE POLITICS OF COOPERATION

However, Briggs went on, another political reason for planetary missions has been coming to the fore, boosted effectively by the Planetary Society, a public interest group of scientists and space buffs promoting planetary missions. It is the politics of cooperation, Briggs said, "of breaking down the barriers between ourselves and the Soviets by getting into exploration activity where they and we use the resources and capabilities that only they and we have."

If that was part of the motivation, Briggs said, "then the kind of resources, five or six billion dollars, whatever it took, would be money well spent. You find the agency, the White House, the Congress having those things in mind when they think about the justification for exploring." Briggs said that he believed that the Soviet interest in the Martian moon Phobos was to determine if it could be used as a base for exploring Mars with a manned program. "I would be surprised," he said, "if there isn't some cooperation with the Soviets on sample return missions."

There appeared to be a consensus among American observers that the Soviet's main concern with Phobos was its potential as a Mars base and source of water. Orbiting Mars at a distance of only 6000 miles from the surface, the larger of the two moons would offer a staging area and base for exploring the planet.

Aaron noted the advantage of supplying a space ship with hydrogen and oxygen propellant from Phobos water rather than drilling for it on Mars (through permafrost or polar ice) with a gravity well 0.38 that of Earth. "Certainly there would be more resources on the Mars surface than there would be on Phobos," Aaron said. "The advantage that Phobos has, if there is water there, is you're not in the gravity well of Mars and the resources become much more transportable. If you start exporting from the Mars system to the Moon, it's easier to get the resources from Phobos to where you might want to use them elsewhere."

Aaron characterized the Soviet Mars program as exciting, although, compared to America's Mariner and Viking missions, the Russians have not done well at Mars so far. It appears very likely that they will get to Phobos first.* "The thing we would like to know is whether or not there is water on Phobos and how easy it is to get," he said.

Will they tell us?

"I think so," Aaron said.

THE HUMAN PROBLEM

Both American and Soviet ambitions for extended manned space flight were circumscribed by the physiological constraint of the effects of weightlessness on the skeletal and neuromusclar systems of the crew. The first American research into these effects was carried out in the *Skylab* space station in 1973–1974. Nine astronauts had spent 171 days, 13 hours, and 14 minutes aboard the space station in increments of 28, 59, and 84 days. The medical results of the three missions showed that although crews adapted successfully to the weightless environment (in some cases after an initial bout of nausea and vomiting, termed *space adaptation syndrome*), the problem of readaptation to Earth gravity was eased considerably by in-flight exercise.

The final medical report stated that the third crew that stayed longest (84 days) and performed the most regular exercise returned in the best physical condition and made the quickest readaptation to one g. Skeletal, cardiovascular, and muscle tissue examinations failed to reveal significant deterioration in any of the astronauts. No "irreversible" effects were detected.

During Project Mercury, NASA physicians had expressed concern about calcium loss from bones

*Two Soviet attempts to reach Phobos failed in 1988 and 1989 to maintain radio contact with Earth.

and muscle atrophy. The *Skylab* studies reported such loss was less than expected. No loss was detected in the forearm bones of any crewman, and calcium depletion in the heel bones was similar to losses noted during bedrest studies. Members of all three crews lost weight (three pounds or less per man). Caloric requirements, however, appeared to be the same on orbit as on the ground.

On the Apollo missions, loss of red blood cell mass had been observed, and this occurred on *Skylab*. However, it lessened with each flight, from 14 percent mean loss for the 28-day crew to 12.3 percent for the 59-day crew and 6.8 percent for the 84-day crew. Plasma loss, however, increased per mission, ranging from 10 percent for all three crewmen on the 59-day flight to 13, 16, and 19 percent among crewmen on the 84-day mission. Recovery of red cell mass was reported to be more rapid among second- and third-mission crews.

Muscle function was better with each mission, accompanying an increase in exercise time from 30 minutes a day on the 28-day flight to one hour on the second flight and 90 minutes a day on the third. The studies found an increase in weight and a reduction in chest and stomach circumference. In each person there was a shift of blood and tissue fluids during flight from the legs to the upper body.

The medical evaluation concluded, "data analyzed to date (1974) and the observations made during and after the flight do not indicate any medical constraints for continued extension of man's duration in space. . . . The *Skylab* crewmen demonstrated that man can fully adapt to a weightless environment, perform in an efficient and effective manner and then readapt to the one g environment. A major milestone in manned spaceflight has been accomplished."

Further experiments in human physiological effects of microgravity became the exclusive province of the Soviet Union's 15-year space station development program, with its eventual objective of determining whether a man could endure a year in space flight—long enough to fly to Mars.

This was one of the medical questions that NASA hoped to resolve in the human biology studies planned for the American space station.

"It's a fundamental question," Aaron commented. Before being posted to Washington to head the Office of Exploration, he had been space station manager at Houston. "We don't know whether we can fully solve the human problem without artificial gravity. We may need to provide it for these long trips. We have probably no other alternative than to do the research.

"At the level at which we have been able to study it, artificial gravity space ships appear to be reasonable . . . and can be feasible. There is an impact in doing that, but it doesn't look like it's beyond the capability of doing it."

Aaron admitted he regarded the problem from an engineering viewpoint. "It's probably within our capability to estimate cost," he said, "we don't have very good cost and weight figures yet; but it appears feasible to do . . . to provide artificial gravity."

THE MUSCLE LAB

At the Kennedy Space Center, studies of human adaptation to microgravity and readaptation to 1 g were being carried out by Dr. Paul Buchanan, a veteran of space medical studies who had taken part in the *Skylab* program. He was director of the Kennedy Space Center Biomedical Office. "The work we did on *Skylab* was primarily designed to maintain the cardiovascular, cardiopulmonary system, and it wasn't until later, on the third [crew] mission [*Skylab 4*], I believe, that we became aware from Bill Thornton's work that we were losing muscle mass and were definitely losing lean body mass," Dr. Buchanan recalled.[3]

"We were losing strength in the lower extremities; probably some back strength, although we didn't have the mechanisms for testing at that time. It's doubtful that anything we did worked very well. With the Mini-Gym, a centrifugal clutch arrangement, and the bicycle odometer, we put on a set of bungee or stretch arrangements to try to encourage the crew to do different things to maintain muscular condition. Only one of the nine crewman actually showed any increase in strength. That was [Owen] Garriott who developed an increase in arm strength. And the reason he did, we think, was because he never really did arm exercises before

and he began using the arm exercise equipment as part of this routine workout.

"We wouldn't want to draw any great conclusions there about what we saw in *Skylab* as being very effective. Matter of fact, I think it was not very effective.

"We have tried to look at the data provided by the Soviet Union. If we give it full credence, we'd have to say that their program isn't very effective either. In fact, they tell you that. . . .

"In spite of two hours a day, they are still showing losses of strength of the back, lower extremities and losses in lean body mass; showing other signs of muscle atrophy of the back and lower extremities. And lower trunk muscles.

"One of the Soviet scientists described it as subatrophy. The muscles did not have the tone we would expect a healthy individual walking around to have. So they're having a problem in spite of the fact that they're using a bicycle odometer, they're using a treadmill, they're using a Penguin suit, that is, a garment which has resistance wires and bungee cords sewn into it . . . so that it doesn't matter what you do, you're working against the suit as if you were in a gravitational field. Whether or not they're still using the lower body negative pressure suit (to balance the movement of blood to the upper body) I really don't know.

"But they're not satisfied that they've found an answer. We are not satisfied that they're found the answer and we are certainly not satisfied that we know enough about the kinds of exercise that will be needed or over what period that we will have to do them to prevent muscle atrophy."

The first *Skylab* flight showed that "we can fly for 30 days without anything," Buchanan said. "We will not come back as green jello. . . . The 59-day crew of Bean, Garriott, and Lousma came back showing some signs of muscle loss, orthostatic intolerance, as we expected, but it was not really a critical thing. And within a few days they were back doing most of the things expected of them, although it took them several weeks to get back to doing all the things they wanted to do."

In the 84-day crew, Buchanan said, there were increased changes even though they were asked to work harder at their exercise program in order to

Figure 4.15. Astronaut Owen Garriott flew on the Spacelab I mission in Columbia in 1983 and as scientist-pilot 10 years earlier on Skylab. (NASA).

ameliorate some of the expected changes. The effects were not linear, from 28, to 59, to 84 days.

"So we can fly 90 days. We can use the same exercise program we had in *Skylab* and get away with it. Is it the thing to do? No! It is not the thing to do. The cost to the crews on returning to one gravity is really too great to be routinely acceptable. We can do better than that."

Buchanan turned his consideration to longer mission planning. What's the easiest way to fly to Mars? he asked.

"From the point of view of wear and tear on the crew on a three year transit, artificial gravity. You and I pick up, we get into the spacecraft and we go. We don't worry about the effects of microgravity on the body. We know that we're going to have the same kind of orthostatic tolerance we had on Earth when we get there; we know we'll have strength in the lower extremities. Yes, there are going to be some problems, very much as you would expect on a small ship, but that's no big deal. We'll survive that.

"The life support system is eased considerably. Plants will grow. Earth normal equipment could be taken with us and we wouldn't have to worry about what microgravity does to plant growth. Up is up and down is down. It's an easy fix physiologically. psychologically.

"But mechanically, I don't think it's an easy fix. I think it would be a great engineering exercise. If we have unlimited money and unlimited time, I would like to go ahead and see the engineers solve this problem immediately. I think we have to work very hard on methods of eliminating the effects of this lack of gravity . . . physiological and psychological mechanisms . . . and get on with the program . . . and hope that artificial gravity will catch up with us."

Buchanan said that a program to develop a prescription for exercise was being studied at the Johnson Space Center in the belief that conventional exercise will prevent the kind of atrophy seen in null gravity. How is conventional exercise to be provided in microgravity?

"Can't use bar bells," the physician continued. "In one g if I have a 30-pound bar bell or dumbbell in my hand, gravity is trying it pull it back down.

As I resist that pull, that's an eccentric contraction of the upper extremities. We're of the opinion that people who work hard to maintain conditioning of the muscles and strength ... body builders .., probably benefit as much or more from the eccentric component of a muscular contraction as they do from the concentric component. We need to find out whether eccentric component is as important as we think it is."

The training guides in some of the good texts for physical education programs are not in agreement, he said, adding: "When you start talking about building muscles, you're right on the edge of pure magic ... pure witchcraft. The transition from the expenditure of energy and the chemical conversion in the muscle fiber is still not well understood."

If the eccentric contraction is as important as some experts believe, a resistance force generating machine will have to be included on the space station, Buchanan said. It's easy to use a centrifugal clutch mechanism that provides a concentric contraction, but such an apparatus does not cause contraction in the eccentric direction. Buchanan described an electrohydraulic apparatus that could provide both eccentric and concentric contractions, but it has some drawbacks. He said it uses a lot of power and hydraulic fluid that would be a potential contaminant in a space station. Work was being done on a double-spring-loaded system, he went on. But some of the more elaborate muscle-building machines were found to be more complicated than the simple centrifugal clutch that was flown on *Skylab,* he said. Before condemning the space station program to a large expense, he said, "we're got to be ready to tell the system managers that you have to do this ... because ... but we can't do that."

There may be another option, however. Stimulation of the muscles by a weak electric current passing through electrodes attached to the skin has been shown to have beneficial effects without inconveniencing the patient to any great extent. "Several years ago," Buchanan related, "in Southern California, there was an attempt to treat a young lady suffering from scoliosis, curvature of the spine. Transcutaneous electrodes connected to an electrostimulation machine caused the muscles of the

back on the convex side of the curvature to contract. Periodically, this was done while she was resting. They found that they could even do it at night, as the girl slept, and it produced an extremely beneficial effect in the scoliosis." However, if the treatment was discontinued for a month or two, he went on, eventually the muscles would weaken and the curvature would return. But as long as the treatment was used, the muscles tightened on the convex side and pulled the spine into vertical alignment.

Buchanan said that in reading an article about the treatment he was impressed by the report that the patient could readily tolerate the electrical stimulation. The treatment, electromyostimulation (EMS), was used experimentally during a bed rest study of muscle atrophy at NASA's Ames Research Center, Mountain View, California. The bed rest simulated the environment of microgravity in orbital flight. Eleven volunteers stayed in bed 24 hours a day for 30 days during the summer of 1987. The electromyostimulation treatment was applied to three of them. The data appeared favorable, it was reported, but not conclusive. Buchanan, a principal investigator for the bed rest experiment, said that he was impressed by the fact that volunteers, three young women, were able to sleep during the treatment. Electrostimulation had to be fairly strong in order to "recruit" muscle fiber, he said.

"It said to me: look, we might be able to use this on the crews; while they're eating, during off time, while writing letters home or on the videophone," he said. "If we don't come up with a good system, we're probably condemned to something that the managers and crews don't want ... and the cosmonauts don't want .. a mandatory one to two hours a day exercise period. That's expensive time.

"If I could put them into a battery power source that would stimulate the muscles of the lower extremities and lower back, that they could tolerate while they rested, or played pinochle, I could hopefully cause enough stimulation in those muscles to keep that muscle tone up, to prevent the atrophy. And make sure they got back in one g with the right kind of strength."

Buchanan considered the question of whether research in muscular dystrophy and associated neuromuscular disorders shed any light on the cause of muscle tissue wastage resulting from long term space flight. "We haven't found anything in their literature that we think is applicable to our problem," he said. "So far, they haven't contributed much to us and we haven't contributed much to them. We hope that will change." However, he said the possibility of research in neuromuscular disease based on observation of tissue changes in microgravity "is really great." "The ability to look at muscular changes in both normal and abnormal or diseased people in that environment (microgravity) and see the relative effect of gravity would be extremely beneficial. And what about arthritis? What happens to an arthritic person in low gravity? What happens at the joint level? Do they improve or not? They could stay for 30, 60, 90 days in the microgravity environment and hopefully feel great and be studied."

Inevitably, the problems of muscle atrophy and bone demineralization in space flight would open up a new horizon in medical research, the clinical facility for which would be the space station.

VIEWS ACROSS THE BRIDGE

On December 21, 1987, the Soviet Union commenced a year-long study of effects of a 12-month orbital mission on a crew with the launch of Vladimir Titov, 40, pilot, and Musa Manarov, 31, flight engineer, from the Baikonur space drome to the *Mir* space station. It was expected that the physical condition in which the cosmonauts managed to survive a 365-day mission would tell the Soviet space establishment whether cosmonauts could survive a journey to Mars, and back, with the application of countermeasures the Soviets were reportedly developing to reduce the physiological effects of long-term microgravity.

In *Mir* the two relieved a two-man crew consisting of Yuri Romanenko and Alexander Alexandrov. Romanenko, 43, had been in *Mir* since February 6 and Alexandrov, since July. When they returned to Earth December 30 in the *Soyuz TM-3* spacecraft, Romanenko brought with him a new world record of 326 days in orbit. He was quoted as saying he

felt just fine as he was lifted out of the *Soyuz* descent module, unable to stand or walk, but able to grin and wave at the television cameras. Preliminary data showed that he had grown one-half inch taller on the mission. On March 6, 1988, Romanenko told a news conference in Moscow he was recovering well. The prospect of a manned Mars landing early in the 21st century gave the problem of microgravity physiology more than academic significance. The Soviets were clearly moving toward the mission and in effect had begun training for it. Without a space station, American studies of the effects of long-term space flight would be forced to borrow from the Russian experience until the American station was operational in the mid- or late 1990s.

However, a unique exchange of medical and scientific views on the physiological effects of weightlessness took place July 18, 1987, between Soviet scientists in Moscow and American scientists attending a Planetary Society meeting in Boulder, Colorado. It was one of several topics discussed in the exchange, carried live on television via satellite.

At Boulder, Dr. Paul C. Rambaut, who had taken part in studies of mineral and metabolic effects on *Skylab* astronauts in 1973–1974 observed that prolonged exposure to weightlessness had remained a major problem for both space-faring powers. Addressing his Soviet counterparts, he said:

Your countermeasures that are used on *Salyut* and *Mir* seem to be fairly effective in the sense that the loss of skeletal mass in orbit is to be a lesser degree than what we see in simulated conditions on Earth— long-term bed rest, for instance. Yet those countermeasures are clearly not completely effective. At the end of your 150-, 210-, 185-, 237-day missions, you do report some loss in bone mass, particularly in the heel. What has been the experience in terms of the recovery of bone mass months after they return from the first long-term flight? [*]

From Moscow, Arkady Ushakov of the Institute of Medical/Biological Problems responded:

[*] Dr. Rambaut had stated at a Skylab Life Sciences Symposium August 27–29, 1974, that, "It was concluded that capable musculoskeletal function will be threatened during prolonged space flights of one and a half to three years unless protective measure can be developed."

Our investigations which all show the same results as obtained from the *Skylab* mission of 84 days allow us to conclude that human beings adapt themselves very well to weightlessness. But we cannot allow complete adaptation because the human being must return to Earth such as he was before the flight without having fully adapted to weightlessness. . . . In space flight many human systems undergo changes: cardiovascular system, muscular system, locomotive function, the vestibular function, metabolism, kidneys, the system of retrophoresis, the system of red blood cells and others, including the nervous system. All of these systems are subject to certain changes during spaceflight. As a result obtained on long cosmic flights, we have developed a preventive (prophylactic) method. . . . This has allowed us to conduct flights of such long duration. Today, in the Soviet Union, based on the long length of our flights, we have concluded that these are sufficient means to counteract the effect of weightlessness and not to allow the human organism to adapt itself fully to the conditions of weightlessness. This also covers the problem of calcium or the bone mass loss which was just asked about by Dr. Rambaut. I must tell you that the experience we've obtained allows us to speak out with a great measure of confidence that the time frame of the flights is acceptable and we do not feel that we should be worried very much about bad results regarding effects of weightlessness on human systems. I could with confidence guarantee you that all our cosmonauts are in very good condition and the systems that are affected most by weightlessness do recover within certain periods. There are no effects that are irreversible and we can with a great deal of confidence extrapolate these 237 days to one year. . . . The medical-biological problems and the physiological problems in weightlessness are being resolved with a great degree of success and in my opinion will add support to an optimistic opinion on the problem of manned flight to Mars.

Dr. John Billingham, head of life sciences at NASA's Ames Research Center, asked about the countermeasures Ushakov referred to. "My question is whether or not the countermeasures you have developed will in fact allow successful performance, successful work and full health on the Martian surface—or whether indeed one needs to add additional countermeasures such as on-board centrifuges or even think about artificial gravity?"

Ushakov replied:

We would like to do additional studies on the necessity of taking measures to create artificial g. because the problem connected with the exit of man onto the surface of Mars after a prolonged flight is a very serious problem. It only partially reflects the data that we have so far obtained from prolonged flights. But taking into consideration that his stay on the Martian surface will not be long and that he then will have to return to Earth, the sequence of staying on the Martian surface followed by the return flight to Earth would raise a great number of very important dilemmas involving the whole project of sending a human being to Mars.

Dr. Evgeny Ilyn, Institute of Medical/Biological Problems, said that America had participated in artificial gravity experiments aboard Soviet satellites. He recalled that Konstantin Tsiolkovsky, the Russian schoolteacher who has become known worldwide as the "father" of space flight, presented the concept of creating artificial gravity at the beginning of the twentieth century. Ilyin said:

The first practical step occurred on board the Soviet Cosmos 782 in 1975. On board there was a centrifuge on which there were placed a number of different life systems and together with our American colleagues we conducted a study of the effects of artificial gravity. We continued experimenting in 1977 on board *Cosmos 936;* in 1979, on board *Cosmos 1129,* and perhaps you know that in all these experiments Americans participated.

For the first time we could experimentally prove that the biological effectiveness of an artificial gravity is identical to our natural Earth gravity. It shows that in principle artificial gravity can be used as one of the means of supporting an optimal level of physical fitness of the cosmonaut . .

The conclusion that artificial gravity would be the equivalent of natural gravity in long-term space flight had important implications, whether or not it was shared by American life scientists. In the view expressed by Ilyin, there was no doubt about its salutary effects, and if accepted by American physiologists, artificial gravity inevitably would become a required engineering design parameter of manned interplanetary space ships. There would be a facility to test this conclusion—the U.S. space station, parts or all of which could be provided with artificial gravity. The extra cost would be a factor, but if American space policy framed a manned trip to Mars and back, it appears reasonable to expect that it would also insist on maintaining the crew in good physical condition. In that case, the provision of artificial gravity would be an essential feature of the life support system.

Dr. Carl Sagan, Cornell University astrophysicist who heads the Planetary Society, asked Dr. Oleg Gazenko of the Institute of Medical/Biological Problems, a pioneer in Soviet space medicine, for his opinion on the limits of long-duration flight in microgravity.

"Perhaps because personally I don't intend to fly to Mars, I am very optimistic and hopeful that men will be able to do that," Gazenko said.

If we are to talk about to what extent our knowledge is sufficient today in order to create a scientific basis for a serious reply to this question, I must say that the book of our knowledge is so far still full of empty pages, in spite of the fact that people have achieved successes in conquering outer space. Nevertheless, the volume of our knowledge is still insufficient in order to fully provide an affirmative answer to the question of whether man will be able to reach Mars. I am sure that the time we have at our disposal prior to actually taking on the task of a manned Mars flight is sufficient in order to obtain this necessary knowledge.

Cosmonauts Titov and Manarov returned to Earth's one g environment December 21, 1988 one year after their orbital marathon aboard *Mir* began. Both were reported to be able to walk after their Soyuz TM-6 reentry capsule soft landed in Kazakhstan.

They told a post landing briefing at Star City December 23 that they were feeling well. They said that they were able to take extended walks in the cosmonauts' compound.

These reports were confirmed by a French Space Agency (CNES) flight surgeon, Dr. Bernard Comet. He said that he was amazed at how quickly Titov and Manarov were able to re-adapt to one g. Soviet specialists credited a program of exercise aboard Mir and strong psychological support for the rapid re-adaptation to normal gravity.[3]

FIVE

THE NEW WORLD

THE Moon is the New World of our time. As it became technologically accessible in 1969, the initial period of lunar exploration transformed the perception of it as a heavenly body to one of an annex of the Earth. It became a prospective source of new wealth in the form of matter and energy for Earth's teeming population. From an evolutionary viewpoint, the settlement and development of this New World—for which detailed plans already exist—are inevitable in the process of human expansion. The process, anthropologists tell us, began for our species in Africa less than a quarter of a million years ago. Since the continental glaciers retreated about 12,000 years ago, human expansion over the planet has been well marked. It was completed in the first half of this century with the continuous occupation of Antarctica.

The commencement of space flight in the second half of this century provides the means of human expansion beyond the Earth, a process of continuing evolutionary scale, analogous perhaps to the emergence of life from the sea. The Moon in this context becomes the first land to be visited by humans in the cosmic sea. However, the return to the Moon has lacked the political and competitive motivations that energized Project Apollo in the first decade of the space age. In that period the development of space flight in America was a reactive process. It responded to external pressures, principally Soviet competition. Apollo was an ad hoc program designed to counter Soviet advances in manned space flight by leap-frogging over them to land men on the Moon.

The scientific discoveries of the six lunar landings and preparatory reconnaissance were politically secondary to the main objective of being first to put men on the Moon.

Once the *Eagle* had landed, the scientific program that revolutionized planetary science was cut back. There was no political reason to continue Project Apollo or its costly transportation system. With the end of Apollo lunar exploration in 1972, the Nixon administration elected to counter the Soviet manned space program by developing the reusable space shuttle. Fifteen years later, the Reagan administration responded to Soviet space station development with a long-sought authorization to build a permanent American space station. Historically, intraspecific competition has been the prime motivation of exploration and expansion by the West since the fifteenth century. But after Apollo it was no longer viable as an energizer for continuing lunar exploration. Except for the prospective launch of the Lunar Geoscience Orbiter for another look, the Moon was on "hold" so far NASA planning went for the balance of the twentieth century.

But a ferment for a return to the Moon was growing in private and government sectors of the space community. It had found expression in the reorganization of NASA headquarters after the *Challenger* accident with the establishment of an Office of Exploration and the Division of Solar System Exploration. It was buttressed administratively by the new presidential goal of expanding the human presence beyond Earth in the Solar System.

The movement to go back to the Moon and explore it rationally as a New World for mankind was intellectual, not political. There was precedent for that approach in the scientific exploration of Antarctica by the United States Antarctic Research Program. The analogy does not bear close comparison, for neither the scientific nor economic potential of these frontiers is closely comparable. One is the last frontier on Earth; the other, the first beyond the Earth. Their material resources are vast, but different. Still, this does not account for their exploration, which expresses the same evolutionary process.

There was a compulsive quality about the push in the space community for the return to the Moon. The late Krafft Ehricke, one of the leading theoreticians of the von Braun team, called it the "extraterrestrial imperative."[1] The term referred to an innate drive in human consciousness to expand the human presence beyond Earth. It was expressed in President Reagan's declaration of national space policy in 1988. The idea that man's fate is to venture outward into the Solar System and beyond has become the official view of the United States.

FROM FANTASY TO PLANNING

Frequently overlooked, the Nixon Space Task Group report of 1969 sets forth the early development of human habitation on the Moon. It was prepared by NASA (as "America's Next Decades in Space") and predicted increasing astronaut stay time on the Moon with some accuracy. The *Apollo 11* lunar module spent 21 hours, 36 minutes, and 21 seconds at Tranquility Base, and Neil Armstrong and Edwin E. Aldrin, Jr., spent 2 hours, 31 minutes, and 40 seconds roaming the surface.

From that start, stay time was extended to 31 hours, 31 minutes, and 12 seconds and extravehicular activity on the surface to 7 hours, 45 minutes, and 15 seconds on *Apollo 12*. Stay time reached 74 hours, 59 minutes, and 40 seconds and extravehicular activity 22 hours, 3 minutes, and 57 seconds on *Apollo 17*, the sixth and final manned landing. The range of the explorers began with 1 km at Tranquility Base and reached 33.8 km (21 mi) at Taurus-Littrow, the landing site of *Apollo 17*.

The success of the landing program in extending surface stay time by astronauts suggested to the Task Group that it would be feasible to plan a lunar surface base by 1978, with a station in lunar orbit two years earlier.

The development of the base was set out in the Task Group report in three phases, each dependent on the Saturn 5–Apollo transportation system. The orbital station would be placed in lunar polar orbit. From there crews could descend in a Lunar Module to any part of the Moon for missions on the surface of up to 14 days. In phase two, construction supplies and roving vehicles would be transported to the Moon. Construction of the base would be carried out in phase three.

The base project would require the development of a two-staged Orbital Transfer Vehicle and a reusable lander. It was suggested in the report that a space station module could be used for the initial surface shelter. A tug that also could serve as the lunar lander could have a dual role as well. It could be used to move satellites in low Earth orbit and as a fourth stage on the Saturn 5 launch vehicle for direct delivery of heavy payloads from the Kennedy Space Center.

The Space Task Group report marked a transition from fantasy and speculation about a human settlement on the Moon to a reality-based scenario, taking into account what was known about the Moon in 1969. The idea of inhabiting the Moon is as old as Western civilization. An ancient reference to a lunar colony appears in the "True History" of Lucian the Syrian, a scholar living in second-century Athens. The work tells of a trip to the Moon by a ship's crew blown there by an Atlantic storm. The visitors become involved in a war between the people on the Moon and the people on the Sun. In the seventeenth century, Johannes Kepler (1571–1630), the father of orbital mechanics, invoked spirits to waft his travelers to the Moon in a fantasy called "Somnium." The swordsman-poet Cyrano de Bergerac (1619–1655) evaporated his people to the Moon with the morning dew in "Voyages to the Moon and Sun." Travel to the Moon is not a recent concept.

The first published discussion of a lunar colony is attributed to Bishop John Wilkins, circa 1638, in "A Discourse Concerning a New World and An-

other Planet."[2] The first American account of a lunar trip was "A Voyage to the Moon" by Joseph Atterly. He devised a spacecraft coated with anti-gravity material. About the same time, 1865, Jules Verne launched his lunar project from Florida with 400,000 pounds of guncotton as propellant from a huge cannon. In 1900 H. G. Wells used an anti-gravity shield to put "The First Men on the Moon." In the twentieth century, lunar travel acquired the trappings of mathematics and science, although the fantasy aspects persisted well into the century. The idea of the Moon as a planet, possibly inhabited, possibly capable of sustaining life, was a staple of fantasy and science fiction for generations.

THE LEGACY OF APOLLO

By the end of 1972 the observations of the astronauts, the data from the instruments they set up on the lunar surface, and the 833.9 lb of rocks and fine soil they brought back to the Lunar Receiving Laboratory at Houston established the Moon as a partially evolved planet. Its resources were vast indeed, ranging from aluminum and oxygen to thorium and uranium. In the absence of a sensible atmosphere, the lunar environment provided a hard vacuum for an array of industrial processes and experiments.

Oxygen was plentiful, bound up in the rocks, and could be readily obtained by heating. This became known early. Later, hydrogen deposits from the solar wind were detected, and with it, the rare isotope helium 3, as mentioned earlier. The technology for exploiting a fusion process using helium 3 remained to be developed, but the extraction of oxygen and hydrogen for life support, electrical power and rocket propellant appeared to be a near-term prospect. Beyond the scientific discovery of the Moon as a planet, its potential for industrial development was worth many times the $24 billion investment in Apollo.

A series of lunar studies funded by federal grants were made in the early 1960s.[3] Those done before the Ranger, Surveyor, and Lunar Orbiter reconnaissance programs made hard data available projected a number of misconceptions. One was a study of how petroleum exploration methods could be applied to the Moon. Another, in 1962, expressed a generally widespread conjecture that lunar craters were of volcanic origin and that water existed on the Moon. The volcanic versus the impact origin of lunar craters persisted as a vigorous debate before and during the landings. One goal of *Apollo 17* was to determine whether volcanism has occurred at the Taurus–Littrow landing site. No evidence of it was found. So far as water is concerned, the Moon may be one of the driest places in the Solar System. No trace of it was found, although scientists have speculated that primordial water may have remained as ice in shadowed regions of the lunar poles. In a lunar polar orbit, the Lunar Geoscience Orbiter will look.

New York University's College of Engineering issued a report in 1963 under a NASA grant on methods of extracting water from a variety of lunar minerals. NASA's Office of Space Sciences that year issued a study of lunar scientific and support operations, landing sites, experiments, and astronaut training for lunar activities. The study became a blueprint for manned activities on the Moon.[4]

At about the same time, two scientists at Bellcom, Inc., proposed that a 1500 pound scientific laboratory be landed unmanned on the Moon for use by a crew to be landed separately in the lunar module. It would be equipped for astronomy and geophysics and serve as a base for surface exploration. The Aerospace Division of the Boeing Company in 1963 devised a program called Lunar Exploration Systems for Apollo. It called for the delivery of 25,000-lb modules to the lunar surface by Saturn 5 vehicles for a base that would house 18 persons. In 1964 Boeing issued a technical report on surface roving vehicles for Apollo exploration. The surface vehicles were electrically powered, wheeled trucks, larger than, but similar in concept to, the Lunar Rover, the electric jeep of *Apollos 15, 16,* and *17*. Boeing became the prime contractor for the Lunar Rover.

Other federally funded studies in 1964 proposed the use of nuclear or solar energy to extract water from lunar rocks and confirmed gaseous emissions from the crater Aristarchus. The Aristarchus emissions had been observed intermittently from Earth by Moon watchers but had been the subject of debate about their origin. The emissions suggested to some investigators that volatile elements and compounds, including water, existed in the interior of the Moon.

As Project Apollo was being developed in the mid-1960s, lunar prospects were broadened to include a scientific investigation of resources on the Moon, a proposal for a survey of lunar landing sites, and methods for establishing a lunar base. NASA's Office of Advanced Research and Technology set up a conference on the technological development of the Moon. New reports on the possibility of discovering lunar water were issued by the Air Force and Texas Instruments, Inc. Preliminary profiles for seven Apollo missions were drawn by the U.S. Geological Survey. An analysis of human factors and of environmental control and life support systems for lunar missions was made by Airesearch Manufacturing Company for the Marshall Space Flight Center.

By 1965, projections of lunar exploration provided detailed engineering descriptions of lunar bases. Studies by General Dynamics analyzed shelter concepts, maintenance and crew operations, logistics, and costs. Lockheed Missiles and Space Company issued a report on methods for emplacing modules and a 100 kilowatt nuclear power reactor at lunar base site. These studies were funded by NASA under the umbrella of Lunar Exploration Systems for Apollo.

THE SUMMER CONFERENCES

The list of federally funded lunar base and exploration studies in 1965 includes a NASA-sponsored summer conference on lunar exploration and science at Falmouth, Massachusetts. At Woods Hole that summer the Space Science Board of the National Academy of Sciences sponsored a more general conference on "Space Science: Directions for the Future." Not since the International Geophysical Year of 1957–1958 had the scientific community in government, industry, and academia been so busy devising programs of scientific exploration as it was in anticipation of Apollo. A new world was opening for investigation; the ships were getting ready to sail. North American Aviation issued a report on scientific investigations that could be made from a lunar base; the Boeing Company issued a proposal for a two-man surface vehicle, MOLAB that would support a fourteen-day sortie on the lunar surface with a radius of 50 mi. Bendix issued a list of MOLAB experiments. Boeing also had an alternate proposal for a shelter consisting of the Lunar Module, with a Rover.

In 1966 Texaco Experiment, Inc., produced a study of lunar subsurface probes for a newly instituted Apollo Applications Program (AAP). Bendix described experiments for a 14-day mission from a Lunar Module, with a "Taxi" vehicle. The staff of the House of Representatives Subcommittee on NASA Oversight of the Committee on Science and Astronautics issued a report on Future National Space Objectives. It suggested that the lunar surface could provide an operational base for the scientific exploration of the Solar System. The report dated July 26, 1966, estimated that a permanent lunar base could be established by 1970, using the Saturn 5–Apollo transportation system. It would be a function of the Apollo Applications Program and could be carried out at a cost of $3.4 billion a year.

The subcommittee projections characterized the lunar base as a step in an extended manned program leading to a landing on Mars on the early 1980s. The report assumed the existence of an Earth orbital assembly station and launch platform. It would be desirable to test prototypes of planetary bases on the Moon, the report said.

The subcommittee recommended a maximum effort to reach the Moon, one that would emphasize the U.S. presence in space. The report called for a "prestige program" that would establish orbital stations by 1972, a lunar base by 1974, manned fly-bys of Venus and Mars in the early 1970s, and an operational Earth Orbital Research Facility and Applications Station in the late 1970s. These programs would cost about $6 billion annually for manned space operations within five years.

This far-ranging report of the House subcommittee anticipated the Presidential Space Task Group report of 1969. It outlined a future space policy that

has been expressed repeatedly since 1966: in NASA's 1969 report, "America's Next Decades in Space," the basis for the Space Task Group Report; in similar studies during the 1970s; in the report of the National Space Commission of 1986; and in the Sally Ride Report of 1987.

It is not that the government of the United States has lacked a space policy since Apollo; it has had one for more than two decades without an overall authorization to implement it.

In 1967 the President's Science Advisory Committee issued a report on "Space Programs in the Post-Apollo Period." It suggested that after the second or third lunar landing, dual Saturn 5–Apollo launches extend the surface exploration plan to seven and fourteen days. One vehicle system would deposit a shelter, supplies, and a roving vehicle at a preselected landing site. Another would bring the crew. Although the PSAC report did not recommend a manned landing on Mars, it suggested that "one of the many interesting factors that could serve to produce an accelerated pace (in the space program) would be adoption of a specific mission objective, for example, a manned mission to Mars by some fixed date, say 1985."

During 1967–1968 a list of regional lunar exploration projects was proposed. By 1969 the NASA Science and Technology Advisory Committee for Manned Space Flight issued a report recommending for the first time the development of the space shuttle and a lunar base.[5] Graduate students at the Air Force Institute of Technology drew up a report on concepts for a lunar astronomical observatory. A NASA working group designed a thermoelectric generator to provide power for lunar surface experiments in the Apollo Lunar Science Experiments Package (ALSEP), such as the seismometers and data transmitters. The generator was designated as SNAP 8, an acronym for System for Nuclear Auxiliary Power. It produced electricity from the decay of plutonium and supplied power for experiments on the Moon starting with the second landing, Apollo 12.

Lockheed Missiles and Space Company made a 1969 study of several propulsion concepts for a nuclear-powered flight system. It favored a rocket stage with 75,000 pounds of thrust from a NERVA engine (Nuclear Energy for Rocket Vehicle Application) which had already been developed. The stage would be reusable and would be operated only in space. Public concern about a launch accident that would release radioactive debris over a wide area barred the launch of nuclear rockets from the ground. The nuclear stage which would function as an Orbital Transfer Vehicle from low Earth orbit to geostationary orbit or lunar orbit would be fueled from a shuttle with a cargo bay 15 feet in diameter and 60 feet long. This 1969 study anticipated the design of an actual shuttle cargo bay three years before its dimensions were established.

In mid-1970 a new approach to lunar and planetary missions using the combination of the shuttle, the Saturn 5, and the nuclear propulsion stage was outlined in a NASA Technical Memorandum by Georg von Tiesenhausen and Terry H. Sharpe at the Marshall Space Flight Center. It was ambitious, envisioning space stations in low Earth orbit, in a low lunar orbit and manned bases on the Moon and Mars. These ideas were projected further in the Space Task Group Report.

The von Tiesenhausen–Sharpe project included a space tug for in-orbit activities and the nuclear-powered orbital transfer vehicle, in addition to a shuttle operating from the ground to low Earth orbit. The study proposed that the shuttle be used to build a modular space station in low Earth orbit of standard modules sized to fit into the shuttle cargo bay—a concept that materialized 15 years later. Moreover, space station modules could be used for the station in lunar orbit, the station in geostationary orbit, and for a lunar surface base.

A doctrine of commonality in extraterrestrial vehicle and base construction has been evolving since then to include a Mars base and interplanetary space ships designed like space stations. The idea would simplify and standardize structures and systems for transports and shelters throughout the Solar System. It would also save money. Major elements of the transportation and space station infrastructure would be deployed initially in low Earth orbit by both the shuttle and the Saturn 5. At least two stages of the Saturn 5 would be required to put the nuclear-powered Orbital Transfer Vehicle (OTV)

into orbit, where it would remain for flights to lunar or geostationary orbits.

The OTV would deposit a standard space station module in lunar polar orbit at an altitude of 60 miles on an early unmanned flight. The crew would be brought to the lunar orbit station on an ensuing OTV mission along with supplies and an Apollo Command and Service Module that would enable the crew to return to Earth. On later unmanned OTV flights, a chemically powered Tug, one or more standard space station modules, and a battery-powered Rover would be delivered to lunar orbit.

The Tug would provide the means of ferrying a crew of three to six persons from the station to the surface for sorties of 14 to 28 days. The 28-day cycle is defined by the period of the Moon's rotation on its axis.[*] The synodic period, from new moon to new moon, is about 29½ days, although as measured against the stars the Moon rotates in 27 and ⅓ Earth days, the sidereal period. In general parlance, the lunar day and lunar night are each about 14 Earth days long from the point of view of an observer on the Moon. Hence, a 14-day sortie probably would be timed to begin at sunrise at the landing site. All the Apollo landing missions were timed to be carried out in daylight.

In 1971 the Space Division of North American Rockwell made a feasibility study of a lunar orbital station in a 60 mile polar orbit. The company also proposed a lunar base using space station modules. The modules would be lifted to low Earth orbit by the shuttle, moved from there to lunar orbit by OTVs, and delivered to the surface of the Moon by landers.

OUTLOOK, 1976

With the end of Apollo, NASA's focus shifted to low Earth orbit. It was as far as the shuttle could go for

[*] The Moon's axial rotation has the same periodicity as its revolution around the Earth. The difference of about 2¾ days between the synodic and sidereal periods is due to Earth moving forward in its path around the sun. The Moon is usually described as orbiting around the Earth, but actually both Earth and Moon revolve around their common center of mass, which is about 1000 miles below the surface of the Earth on the side facing the Moon. This point is called the barycenter. It, rather than the center of the Earth, follows an elliptical path around the sun.

the rest of the century. Still, NASA theoreticians continued to regard the Moon as a major objective. They sought a pragmatic rationale to return there. It envisioned a lunar base from which the Moon could be developed as an industrial annex and trading partner of the Earth. How this could be done was set forth by a NASA study group in the document "Outlook for Space" (1976), referred to earlier. Although designed as a general prospectus of the shape of things to come, the "Outlook" projected a lunar development scenario that suggested a wealth of natural resources in this cold, airless wasteland.

Initially, the "Outlook" surmised that a lunar base would be built of materials brought from Earth and shielded from cosmic radiation by lunar soil. Radiation from the sun and other stars was not a major concern on the Apollo landing missions. The astronauts were protected by their space suits and the Lunar Module, and their stays were short. The missions were flown in quiet sun periods. For long-term occupancy, shelters had to be shielded. This could be done effectively, the "Outlook" reported, by putting the shelter—a space station habitat or laboratory module—in a shallow trench and covering the structures with two or three feet of lunar soil.

Without an atmosphere and a radiation-blocking ozone layer, the surface of the Moon was awash with ionizing radiation. Unshielded terrestrial life could not survive in such a sterile environment. No evidence that life had ever existed there was found. American astronauts were the first life forms. To occupy this world, humans would have to create and maintain an artificial biosphere where crops could be grown in sunny lunar greenhouses and workplaces set up in shelters.

Once the lunar base had been established, lunar resources could be processed to provide structural material, such as glass, aluminum, titanium, and concrete. Oxygen could be extracted from the rocks as mentioned earlier, for life support and rocket propellant. In that period it was believed that in the absence of water, hydrogen would have to be imported from Earth.

The "Outlook" proposed that food could be grown in the soil in a greenhouse. A nuclear power station could be built using lunar thorium converted to uranium 233 in a breeder reactor to produce electricity. Surplus nuclear power generated on the Moon could be exported to Earth by microwave or laser transmission at what the "Outlook" described as "multigigawatt" (billion-Watt) levels.

Known lunar resources, identified by the Apollo and earlier reconnaissance programs, indicated that a lunar colony could achieve a high degrees of self-sufficiency. Agriculture under pressurized domes was expected to flourish with 14 (Earth) days of solar illumination and might be carried on during the long lunar night by artificial light. Experiments at the Lunar Receiving Laboratory, Houston, had demonstrated that terrestrial plants could tolerate lunar soil when it was mixed lightly with terrestrial soil.

The "Outlook" analyzed the possibility of manufacturing solar power satellites as a major lunar industry. It suggested that a photovoltaic power system serving Earth could be worth tens of billions of dollars a year. Moreover, nuclear power systems operating on the Moon would not present the risk of endangering Earth's biosphere.

The far side of the Moon, the side turned away from Earth, would provide a special environment for radio and optical telescopes, well shielded from Earth interference. Low lunar gravity made it feasible to erect large structures and the airless surface offered as clear "seeing" as could be obtained in space plus a stable platform.

LAGRANGIAN POINTS

Although the raw material for manufacturing solar power satellites and other products is available on the Moon, the mineral composition of the ores on the Moon differs from that of terrestrial ores. The "Outlook" mentioned aluminum as an example. On the Moon the primary source of aluminum is anorthosite, instead of bauxite, as on Earth. Lunar iron and titanium would be extracted from a mineral called ilmenite. Some ore recovery processes on Earth require water, which has not been found on the Moon. Consequently, refining would have to be adapted to an anhydrous environment.

The "Outlook" alluded to private studies of the feasibility of creating manufacturing facilities orbiting the Moon at Lagrangian points in space where gravitational forces of the Earth, Moon, and sun are neutralized and a stable orbital position can be maintained.* The advantage of orbital manufacturing, its proponents claim, is a saving in transportation costs, even from the Moon. Instead of boosting an assembled satellite off the Moon, the raw material would be launched to an orbital factory for processing and assembly. It would be thrown from the lunar surface by an electromagnetic catapult to a collecting area and picked up there by a tug and delivered to the orbital factory.

Several electromagnetic catapult designs have been proposed. One described by personnel at NASA's Ames Research Center in 1973 would accelerate a load of several tons over a track to lunar escape velocity. The accelerator consists of a stationary coil that includes the track and a moving coil, managed from a control room. Opposing magnetic fields in the coils produce the acceleration, the moving coil moving along the inside of the stationary coil. When escape velocity is reached, the payload is separated from the carrier and the moving coil is decelerated. The final velocity is 2900 meters (9514 feet) a second, according to the Ames design. The length of track required for the payload to reach this velocity at a constant acceleration of 10 g would be 40 kilometers (24.8 miles).[6]

LOS ALAMOS, 1982

When the shuttle began flying in 1981, it was hailed by many in the space community as the genesis of a new national transportation system that would eventually reach the Moon and planets. The next step projected by NASA, the House space subcommittee (1966), the President's Scientific Advisory Committee (1967), the Space Task Group (1969),

and the "Outlook for Space" (1976) was the space station in low Earth orbit and an Orbital Transfer Vehicle system to extend manned operations to the Moon.

In January 1982 a proposal for a national commitment to build an international research station on the Moon was drafted by Paul W. Keaton and Eric M. Gelfand of the Los Alamos Scientific Laboratory. The laboratory would be self-sustaining and available to scientists from all parts of the world. It would be developed, they said, with a "broadly based infrastructure of stations, vehicles and programs that can be envisioned as a pyramid resting on Earth and reaching the the Moon."

The Los Alamos proposal designated the shuttle as the first element of the infrastructure. The second would be a manned platform in low Earth orbit. Next would be the Orbital Transfer Vehicle designed to carry crew and cargo between the platform and lunar orbit. The final element would be the self-sustaining laboratory on the Moon. "By 2000, the laboratory would be a valuable asset for basic and applied science programs in the natural and social sciences," the proposal said. "After that, the knowledge gained could provide ways to use the Moon's resources and environment." The laboratory could be used also as a springboard for unmanned deep space probes, the proposal added, and a much-discussed space settlement, a prospect developed by studies at Princeton University, the Ames Research Center, Stanford University, and a NASA-sponsored faculty fellowship program in engineering systems design (1975).

Keaton and Gelfand cited a 1968 "Moonlab" study by the Stanford University–Ames Research Center Faculty Workshop in Engineering Systems Design. It described a self-sustaining lunar research station with a mass of 276 tons housing 24 persons. Developing such a station over 15 years would cost $17.4 billion (in 1968 dollars) compared with $25 billion spent on the Apollo program.† The design and engineering of a lunar research station, Keaton and Gelfand said, could lead to the assembly of a large technical base on the Moon. Materials sci-

---

*There are five points in Earth–Moon system space where a satellite or station can remain relatively stationary. They are called the Langrangian points, for Joseph Louis LaGrange (1736–1813), the French mathematician who defined them.

† Estimates of the total cost of Apollo vary from one source to another. The cost of building and operating the Saturn 5–Apollo system for the lunar program is generally reckoned at $24 billion.

ence would benefit from advantages of the Moon's one-sixth gravity field, negligible magnetic field, and high vacuum. "The entire enterprise could provide a rapid increase in scientific and engineering manpower to increase our nation's productivity," they said.

The Los Alamos initiative struck a responsive chord at the Johnson Space Center, Houston, where Michael B. Duke and Wendell W. Mendell of the Division of Planetary Sciences had been working and thinking along the same lines. At the Fourteenth Lunar and Planetary Science Conference in 1983, they reported that their division people had been thinking about the Moon in the context of a general re-examination of research goals and programs in planetary science. Several days later, Duke and Mendell said, they "stumbled" across the Los Alamos proposal. "The parallelism between the Los Alamos arguments and ours was striking, almost eerie," said Mendell. "That was the first of many encounters with groups and individuals mostly outside the space program who see a manned lunar base to be important as the generator of a bow wave in the nation's science and technology innovations."

At the Fourteenth Lunar and Planetary Science Conference, Duke and Mendell had proposed that a new start for a lunar mission be inserted in the NASA budget by the 1985 fiscal year. The establishment of a manned lunar laboratory would "impact every part of NASA's research and development organization," they said.

The Johnson Space Center scientists projected a lunar initiative in three phases. The first phase would acquire scientific and technological information for a realistic appraisal of a lunar program by the 1990s. The second would send unmanned roving vehicles to the lunar surface for site evaluation, civil engineering measurements, and scientific experiments. Automated factories would be set up to commence processing lunar raw material before a permanent manned presence was established. The third phase would establish a permanent manned base.

The first habitat module would be landed where an automated factory was already producing oxygen from the soil. A remotely controlled soil mover would deposit the module in a small depression and cover it with soil for radiation protection. The initial crew, the proposal said, would consist of 12 persons. Additional modules would be emplaced to accommodate added personnel. As research facilities became operational, scientists from all parts of the world would be transported to the facility to perform experiments in astronomy, high-energy physics, geology, and the life sciences. Concluded Mendell: "As life becomes routine and the exotic flavor passes, schoolchildren will wonder why anyone ever doubted that the Moon would be an integral part of our destiny."

1984: THE LUNAR INITIATIVE DEFINED

By the end of 1983, two decades of proposals from NASA, the universities, government laboratories, private laboratories, and industry had led to a near consensus on possible development on the Moon. It was generally anticipated in the space community that an antarctic style lunar base would be the logical sequel to Apollo.

As the shuttle became more or less operational in 1984, the transportation and logistical strategy became evident. The shuttle would provide the initial lift to low Earth orbit where a space station would serve as a transfer node to geostationary or lunar orbit. The energy required to reach either orbit was of the same order. An Orbital Transfer Vehicle designed to move payloads from low Earth orbit (LEO) to geostationary orbit (GEO) could also deliver about the same payload to lunar orbit (LO).

If construction followed the doctrine of commonality, modules designed for the space station would also provide initial shelter for a lunar orbital station and a lunar surface base. That idea had been established for a decade. A lunar module or tug would shuttle crews between lunar orbit and the surface after habitat and laboratory modules had been delivered along with logistical equipment. It did not stretch the imagination to suppose that a series of nuclear-powered OTVs could boost a space station habitat module with crew to Mars, with an Apollo-style lander and Earth return logistics.

When the shuttle approached operational status, as NASA defined it in 1984, President Reagan had committed the nation to build a permanently manned station in low Earth orbit. The much discussed and long planned second step toward human expansion beyond the cradle of the Earth was now in order.

LUNAR BASE VS. SPACE STATION

In 1983 the Division of Policy Research and Analysis of the National Science Foundation called upon a consulting firm, Science Applications, Inc. of McLean, Virginia, and Schaumburg, Illinois, to compare the scientific advantages of a base on the Moon with those of a space station. The consultant considered the relative effectiveness of each facility for investigations in high-energy physics, selenology (or lunar geology), planetary science, and life sciences. The report concluded that a set of lunar base experiments was largely separate and in some instances complementary to research that would be done from a space station.

In astronomy, for example, the lunar base would be more advantageous as a site for an optical telescope. It offered a stable platform for a 100-meter (328-foot) thinned-aperture telescope with an an-

Figure 5.1. This Martin Marietta concept illustrates a reusable upper stage for the space station and the shuttle. It is an orbital transfer vehicle (OTV) for the transfer of cargo between low Earth orbit (LEO) and geosynchronous orbit (GEO) at 22,300 miles altitude. In this view, the OTV extends its parasol-shaped aerobrake to reduce its velocity as it approaches the space station in low Earth orbit. (Martin Marietta, Denver).

THE NEW WORLD

gular resolution at least 10 times better and light-gathering power 100 times greater than observing instruments placed in orbit. Pointing stability would be superior on the Moon as well, the consultant's report stated.

Because background interference would be minimal on the Moon, a lunar far-side observatory would be advantageous for radio astronomy, especially for very long baseline interferometry. Interferometry could increase the observation of distant objects by orders of magnitude, using the Earth and the Moon as the baseline. With this arrangement, the report said, an observer could discern sources of rapid signal time variations in extragalactic radio emissions. It was postulated that a highly sensitive ob-

servatory on the far side of the Moon would be useful in the search for extraterrestrial intelligence as manifested by radio signals.

In the realm of high-energy physics, the consultant reported, the Moon provided a hard-vacuum and low-gravity environment for large atom smashers. The report cited as an example the proposed Desertron, a super nuclear particle accelerator covering a 12-mile track. Lunar soil would provide the construction material, including iron for the magnets. Beyond these advantages the report supported previous studies suggesting that space structures, such as solar power satellites, could be manufactured on the Moon and launched from there at lower cost than that of Earth-based manufacture

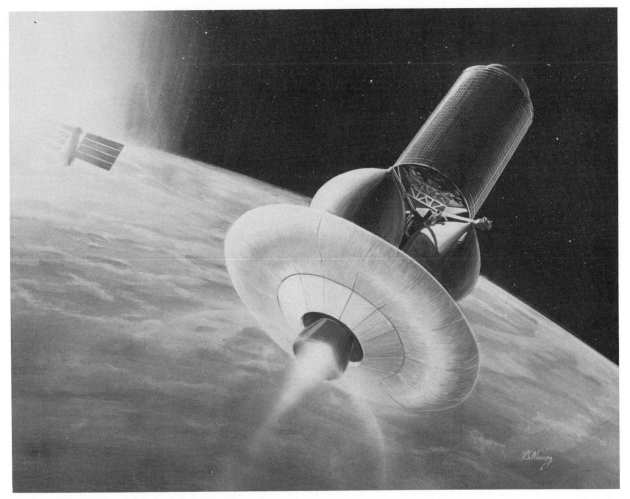

Figure 5.2. With delivery of its payload to a geosynchronous orbit, the orbital transfer vehicle would fire its rocket engine to break out of the high orbit and began descent. The aerobrake is then deployed to decelerate the vehicle as it encounters atmosphere, allowing it to fall into the space station's lower orbit. (Martin Marietta, Denver).

and launching. Escape from the Moon required only 5 percent of the energy required to escape the Earth's deeper well of gravity.

Science Applications devised a lunar base plan that would provide a space station style module for a crew of six to eight persons. A crew of seven would include a commander–pilot, pilot–mechanic, technician–mechanic, physician–scientist, geologist, chemist, and biologist–physician. Because of the likelihood of long stay times at the lunar base, a design providing 2000 cubic feet per person in the base was proposed. In order to realize that much volume per person, the base would require three modules with full life support equipment. Like the station in Earth orbit, the Moon base would consist of habitat, laboratory, and resource modules, all sized to fit the shuttle cargo bay and the Orbital Transfer Vehicle. The volume available to each crew person would be similar to that aboard a Trident class nuclear submarine, the consultant said.

HOME ON THE MOON

The base would require a 100 kilowatt nuclear power generator, surface transport vehicles, a chemical processing plant to recover oxygen from the soil, and a mass mover. The Earth-to-Moon transportation system would consist of three vehicles: a two-staged Orbital Transfer Vehicle, a logistics payload lander, and a Lunar Module.

When the lunar base site was selected, a survey team would set up beacons there. Automated or remotely controlled logistics landers would be directed to the site to deposit base modules and equipment. The consultant estimated that delivery would require five logistics flights.

The three-vehicle transportation system would operate as follows. Launched from a base or space station in low Earth orbit, the two-stage OTV would boost its logistics cargo on a translunar trajectory. The first stage would then separate and return propulsively to the Earth orbit base. As it approached the Moon, the second stage would drop an expendable lander containing the payload and swing around

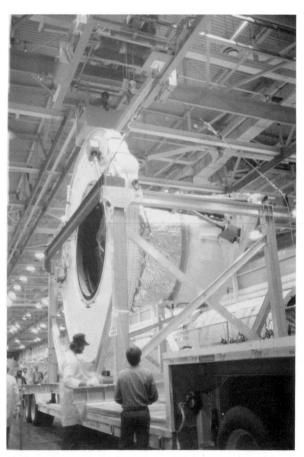

Figure 5.3. For transfers of payloads from low Earth orbit to geosynchronous, lunar or planetary orbits, an upper stage booster has been designed to be carried in the shuttle cargo bay. Shown here at the factory is the pedestal/base of the booster, the Transfer Orbit Stage. (Martin Marietta, Denver).

the Moon to go back to Earth on a free-return trajectory. Applying an aerobraking maneuver to shed lunar return velocity, the second stage would dip into and skip out of the atmosphere to fall into low Earth orbit. The logistics delivery process would be repeated until module and equipment deliveries were completed. The OTV would then be configured to bring the construction crew. This time, the second stage would go into lunar orbit, to allow the crew (in groups of four per trip) to transfer to a waiting Lunar Module. The LM would serve as a lunar shuttle between lunar orbit and the surface. It would be reusable. Later, as base operations reached the stage of crew rotation, a crew returning to Earth would board the OTV passenger mod-

ule from the LM as the newly arriving crew got on to descend to the base.

The Science Applications and Los Alamos studies were among the earliest to describe enabling technology for the occupation of the Moon based on the shuttle, the space station, and the two-staged OTV, all within the state of the art. As far as the scientific advantage of the lunar base compared with the space station was concerned, there was no winner. The lunar base required the space station, but not vice versa. There was a vast difference in cost. The consultant estimated that, compared with $14.2 billion (1984 dollars) for the space station, the lunar base and its transportation infrastructure would cost $54.8 billion (1984). The con-

Figure 5.4. In this artist's sketch, the Transfer Orbit Stage (TOS) is shown as it is boosted out of the base of the stage. (Martin Marietta, Denver).

sultant gave this opinion: "Because plans for a space station are more advanced and its scientific program is broader, it is logical to do the space station first."

Los Alamos, 1984. The first general conference of scientists and engineers on a lunar base was sponsored by NASA and held at the Los Alamos branch of the Institute of Geophysics and Planetary physics of the University of California on April 23–27, 1984. It was directed by Michael B. Duke of NASA's Johnson Space Center and attracted 49 persons from NASA, universities, scientific laboratories, and aerospace companies. Among the participants were Hans Mark, NASA deputy ad-

ministrator, and Harrison H. Schmitt, former astronaut and former senator from New Mexico. Dr. Schmitt, a geologist, was the only scientist to land on the Moon (Apollo 17). Dr. Mark was the former director of the Ames Research Center.

In its report the working group of the conference formulated a rationale for returning to the Moon and setting up a lunar base. "Historically," it said, "the Moon has been an important objective of each new wave of technical capability by the space-faring nations. It is entirely reasonable for us to anticipate that the establishment of a permanent presence in space will extend to a manned outpost on the lunar surface once the enabling technology is in place."

Figure 5.5. A Transfer Orbit Stage is shown lifting a payload such as a Tracking and Data Relay Satellite from low Earth orbit to geosynchronous orbit. (Martin Marietta, Denver).

The group predicted that the space station would change the way society views space and that subsequently Americans would focus on occupying rather than merely visiting space. A permanent base on the Moon would offer the most "robust" combination of factors for long-term growth of the nation's future in space, the working group said. The conferences report cited the potential for developing the Moon as an industrial annex of the Earth, as well as its scientific importance as an outpost of terrestrial civilization.

Summarizing the Moon's industrial possibilities, the report reiterated the thesis that the Moon has ample resources of silicon, iron, aluminum, calcium, titanium, and all other elements except those that are biologically important—carbon, nitrogen, and hydrogen. Later, it was theorized that hydrogen deposited by the solar wind could be collected from the lunar regolith by surface mining.

Iron, aluminum, titanium, silicon, and oxygen could be extracted from the regolith by chemical and electrochemical conversion methods. Oxygen would be an early product. Silicon could be used in photovoltaic (solar) cells. The possibility of collecting hydrogen, nitrogen, and carbon from solar wind fallout was considered, but it was widely assumed at that time that hydrogen and nitrogen would have to be imported from Earth to satisfy life support needs and the production of rocket propellant. Beyond power and processing plants, heavy equipment was required. The working group anticipated that two front-end loaders would be needed for site

Figure 5.6. A Transfer Orbit Stage is depicted here launching the Mars Observer satellite to Mars orbit. (Martin Marietta, Denver).

preparation. This required digging a trench in which to deposit the modules and pushing soil over them for cosmic ray protection.

On the basis of existing transportation technology, the cost of transporting material to the lunar surface was calculated at $10,000 to $15,000 a pound. Heavy lift launchers derived from shuttle boosters might reduce the cost by $3000 to $5000 a pound. In general, two thirds of the cost of the lunar base would be in transportation. The working group calculated that a significant cost reduction might be achieved by developing reusable OTVs using lunar oxygen with terrestrial hydrogen as propellant.

As in the Science Applications analysis for the National Science Foundation, the working group proposed using the OTVs between low Earth orbit and lunar orbit, with an aerobraking return to low Earth orbit. The working group estimated that the cost of the lunar base would be comparable to that of the Apollo program. The group expected that the base could be supported without increasing NASA's historical allocation in terms of percentage of the federal budget.* Participation in the lunar base program by Western Europe and Japan would contribute support.

The working group also considered the effect of four multilateral space treaties that have entered into force in 25 years of space activity but that have not been ratified by the United States. These will be discussed later. Although principles set forth in two of these treaties would restrict development of lunar resources by any single nation, the group stated that "one should not assume that territorial sovereignty will be legally precluded forever." It noted that the concept of functional sovereignty appears to be implicitly legitimate in space law and that states retain rights to exercise jurisdiction and control over their own activities in space. The working group concluded, "In this venture from Earth . . . we can begin a new age in which the horizons of mankind are expanded and aspirations of future generations are fulfilled. We can continue to explore the heavens and for the first time, live there."

* Generally reckoned as about one percent.

THE MOON SINCE APOLLO

Following the 1984 Los Alamos meeting, a more comprehensive conference was held that year in Washington on Lunar Bases and Space Activities of the Twenty-first Century. It led to the formation in NASA of a Lunar Geoscience Working group that reviewed the state of lunar geoscience since Apollo and projected its future course. The group sought to address three questions:

> What is the Moon good for?
> Where did it come from?
> What does it tell us?

In the working group's review (1986), the first question was addressed by Geoffrey A. Briggs, director, Solar System Exploration Division, NASA headquarters. "The Moon is a special world that preserves the early history of the Solar System and it is the key to understanding how all the terrestrial planets formed," he said. The crust of the Moon, the report said, has been differentiated by volcanism and the bombardment of planetoids and meteorites, and the surface has been well preserved since the first few million years of Solar System history. It provides a laboratory for the study of early planet evolution, at least in the inner Solar System.

"For human affairs, establishment of a long term lunar base and utilization of lunar resources are inevitable, if not imminent, in the post-space station era," the report said. The report regarded humanity's return to the Moon as an article of faith. The first step would be a reconnaissance of the Moon's geological and geophysical characteristics by a polar orbiting satellite. It would be launched from Earth in the early 1990s, the report said.

As mentioned earlier, this satellite, the Lunar Geoscience Observer, would chart and make chemical studies of unexplored and unmapped regions, mainly at higher latitudes than those observed by American and Soviet reconnaissance in the 1970s. Of particular interest to future manned activities, the report said, are the lunar polar regions. It reiterated speculation that primordial ice may still exist in the permanently shadowed craters there. The ice would be a source of hydrogen as well as oxygen.

Analysis of the Moon's surface material, the regolith or as members of the working group called it, the "megaregolith," referring to the crust as a whole—has been based largely on samples, and some remote sensing in the 1970s. A total of 833.9 pounds of soil and rocks was brought to Earth by six Apollo landing expeditions from July 20, 1969, to December 19, 1972. Three Soviet sample return flights (automated) picked up a total of about one pound of soil on September 20, 1970; February 14, 1972; and August 9, 1976. The nine samples were taken from the dark colored maria and the light-colored highlands.

These samples were analyzed by international laboratories. In addition, a general picture of the location of heavy metals, especially uranium and thorium, was obtained by remote sensing instruments aboard the *Apollo 15* and *16* Command Modules in lunar orbit. The Lunar Geoscience Group called upon NASA to launch new sample return vehicles and implant automated instruments in the surface that would radio seismic and cosmic ray data to Earth. Seismometers and cosmic ray detectors left at the landing sites by Apollo missions a generation ago have been silent for years.

The principal assessment of the nature and evolution of the Moon that exists today is mostly the product of four years of active Apollo exploration, supplemented by Soviet data. The Apollo data were organized in a series of Lunar Science Conferences in the 1970s, which evolved into annual Lunar and Planetary Science Conferences held at Houston.

During the early assessment of Apollo results, analysis of crater formation on the Moon led to the deduction of a massive bombardment of the inner Solar System starting 600–800 million years after the Moon formed. Scientists viewed the phenomenon as a terminal cataclysm that not only shaped the landscape of the Moon but bombarded Mars, Earth, Venus, and Mercury. It was regarded as a terminal phase of the accretionary process that formed the Moon and the terrestrial planets.

Ages of major basins on the Moon, determined from analysis of rock samples, indicated that the terminal cataclysm lasted about two billion years. The Serenitatus and Imbrium basins appeared to have been punched into the lunar crust by plane-

toid impacts 3.87 and 3.85 billion years ago, respectively. It was suggested by investigators that Mare Imbrium was gouged out by a planetoid the size of the island of Cyprus. The oldest basin dated in the Apollo collection period was Mare Nectaris, formed 3.92 billion years ago. The group noted that other basins appeared to be older than that, but their ages cannot be determined without further surface data. This is an objective of new sample return missions, targeted for suspected ancient basin sites.

There are not enough data to tell how long the terminal cataclysm actually lasted, the group reported. Future sample returns may resolve the question. Investigators regard it as central to the understanding of how the planets formed, how long the terminal cataclysm—the main event of the accretionary process—went on. Meteorites still fall on the Moon and blaze fiery trails as they burn up in the atmosphere of Earth. Accretion is still going on, at a lesser rate. Essentially, this process by which the large bodies sweep up small ones by gravitational attraction is the planet-forming process in our Solar System. It may be universal. This was one of the earliest lessons in cosmogony learned from the scientific exploration of the Moon.

There are more lessons to be learned about the origin and evolution of the planets and their moons. The lunar geoscience group suggested that the lunar surface will contain a record of changes in the sun as well as in the Solar System. Solar observations so far have led to a general conclusion that the sun's activity has been virtually constant for the last four billion years, the group said. But lunar sample studies hint that this conclusion may be premature. Evidence of periodic variation in the ratio of two isotopes of nitrogen, N 15 and N 14, has been found in the solar wind deposits on returned lunar samples. The lunar surface is so quiet that solar wind elements may accumulate there undisturbed for eons. The science group considered this discovery as an indication that something was happening in the sun to affect the isotopic composition of its atmosphere.

An analogous discovery was made in analyzing the isotopes of xenon in the lunar samples. Some of the older rocks contained products of the radioactive decay of xenon (I 129 and Pu 244) that are now extinct on Earth. The group surmised that these short-lived nuclear species had been trapped early in the Moon's formation. The isotopic compositions of solar wind argon and krypton have been virtually constant over the lifetime of the Moon, the rocks show, while helium, nitrogen, and possibly neon have gradually gained heavier isotopes. These variations in solar wind fallout betray variations in the energy processes of the sun, the group noted, surmising they were limited to the light elements. But the group considered that these discoveries in the rocks and dirt of the Moon may lead "to a reappraisal of our ideas about solar evolution."

Billions of years of pounding by planetoids, asteroids, meteorites, and unconsolidated debris smashed the primordial surface of the Moon to rubble. Under continuing impact pressure, rubble was consolidated into aggregates of rock called *breccia* (a term for aggregate formations on Earth). The lunar science review group confirmed earlier estimates that the breccias form the bulk of the surface in the lunar highlands (or terrae), the predominantly light-colored area of the Moon. Drilling efforts in the highlands have indicated that the highland regolith is about one kilometer (0.62 miles) thick. More recent studies suggest greater depth, but determination awaits future missions.

The lunar maria, the dark areas of the Moon which the ancients called seas and oceans, are another story in the history of the Moon. They were formed by two events: planetoid impacts that gouged out the basins and the inflow of molten rock surging up from the interior during epochs of volcanism. As on Earth, lunar volcanism has been generated by internal heating from the decay of radioactive elements. These appear to be as abundant on the Moon as on Earth, although precise surveys have not been carried out.

Lunar volcanism took a different form than the volcanic activity seen on Earth and Mars. There are no towering volcanoes on the Moon. Rather, it appears that molten rock erupted from low-lying crustal fissures and filled the basins. The lava cooled, forming a dark, surface rock called basalt, an igneous rock well known on Earth. (It forms the palisades of the Hudson River and fills ocean basins on Earth as on the Moon.)

In the lunar maria, the basaltic fill has been estimated from 0.6 miles at the edge of some basins to about 3 miles in depth at the center. In their report of recent studies, the lunar science group suggested that the internal heating epoch that produced the lava lasted longer than the two billion years estimated in the 1970s. It may have continued from 1 to 4.2 billion years, the group said. The volume of lava produced by this period of heating has been a topic of conjecture. It has ranged from a magma sea 120 miles deep to the melting of the whole Moon. The actual extent of the melt will not be known until a global survey is made, possibly by the Lunar Geoscience Orbiter.

A problem that has nagged lunar geologists since the Apollo era is the abundance and distribution of a composite soil called KREEP. The term is an acronym for potassium (K), rare earth elements (REE), and phosphorus (P). Initially found by the Apollo 12 and 14 missions in Oceanus Procellarum, KREEP has been traced to the Fra Mauro formation and debris from the Aristarchus and Archimedes craters. Associated with KREEP has been an abundance of thorium, at ten times its global surface concentration of 1.3 parts per million.

KREEP seems to be a very ancient mix, predating 3.9 billion year old breccias. It may have been formed 4.35 billion years ago and extruded from the deep interior to the surface by volcanism. The current idea, the group reported, is that KREEP represents the final product of crystallization, of what may have been the great ocean of magma.

THE MANTLE AND CORE

Like Earth, the Moon has a mantle and core below the crust. In the 1970s the crust was supposed to be up to 60 miles thick from seismic data, three to four times thicker than Earth's crust. The lunar science group believed this estimate might change with further seismic measurements. The upper mantle was estimated 300 miles deep from seismic data.

The major uncertainty about the structure of the Moon is the core. Was it comparable to Earth's core of molten iron? The nature of the lunar core would have a strong bearing on the origin of the Moon, an unsettled problem. The existence of an iron–nickel core in the Moon has been nearly but not entirely ruled out because of the Moon low mean density.

Lunar sample analyses showed that the Moon is depleted in iron. The finding is consistent with the Moon's lower density, 3.34 grams per cubic centimeter, compared with Earth's average density of 5.5 grams per cubic centimeter. The possibility of an iron core with a diameter one fifth the radius of the Moon (about 216 miles) was proposed by a British geophysicist, Keith Runcorn of the University of Newcastle upon Tyne. This would be compatible, he said, with the Moon's density and its moment of inertia.[7] The likelihood of an iron silicate core was suggested by other investigators.

Of course, the question of the Moon's core and density bore on the most persistent mystery in lunar studies, its origin. Had it fissioned from Earth during early planetary formation, as proposed in 1883 by Sir George B. Darwin? Had the Moon formed separately and been captured by Earth? Were Earth and Moon a double planet system, a product of binary accretion?

These questions were hotly debated during the Apollo lunar conferences and remain unresolved. Reviewing them, the Lunar Geoscience Working Group did not express a position but commented that if the Moon had an iron core, it must be a small one, in view of its iron depletion. "This possibility has been a constraint on theories of lunar origin," the group said, "suggesting the selective accretion of either iron poor planetesimals under the influence of a nearby Earth or forming mainly from terrestrial mantle material." The latter case would support the fission hypothesis.

The group reviewed two lines of evidence relating to a metal core, seismic and magnetic. Automated seismometers set up by the Apollo 11 and 12 expeditions transmitted seismic wave data generated by the impacts of meteorites and the crash of the empty Saturn 5 third stage on the Moon.

The group recalled that an early interpretation of the arrival of a single seismic P-wave (pressure wave) from the impact of a meteoroid on the far side of the Moon was "consistent" with the presence of an iron-rich core of 170 to 360 kilometers (105.6 to 223.7 miles) radius. In a later evaluation of this result, the group said, no reliable seismic evidence was found for or against a metallic core with a radius of less than 500 kilometers (310.6 miles).

Scientists have attempted to determine the existence of a core by electromagnetic soundings, laser ranging measurements of the Moon's libration that might indicate a fluid core, studies of the Moon's moment of inertia and paleomagnetic data. A puzzling feature of this research has been the discovery of residual magnetism in the samples and in magnetometer results from instruments Apollo astronauts placed on the surface. During the pre-Apollo reconnaissance of the Moon no magnetic field was detected and investigators concluded that the Moon was magnetically dead. It was not always so. The crustal rocks had been magnetized in the past. Dated soil samples revealed that magnetic fields had been generated on the Moon 3.8 to 3.6 billion years ago. By analogy with Earth, this ancient or paleomagnetism was interpreted as having been generated, dynamo style, by a rotating Moon with an iron core. There seemed to be no other explanation. Presumably, the dynamo process faded as the Moon's axial rotation slowed and became locked with its 27 and 1/3 day revolution around the Earth. But the dynamo hypothesis is circumstantial; the remanent magnetism does not offer irrefutable proof and remains anomalous.

An important clue to the origin of the Moon is its bulk composition. How does it compare with Earth's mantle? There are similarities and differences. The most striking similarity was found in the composition of oxygen isotopes. The group said this comparison comes nearest to proving a "close genetic relationship." On the other hand, the Moon is depleted in volatile elements as well as iron, and samples show higher concentrations of refractory (heavy) elements, such as thorium, uranium, and rare earth elements, than those known on Earth. If the Moon was torn from Earth's mantle, these differences are hard to explain.

The Lunar Science Working Group concluded that all three theories of the Moon's origin—fission, formation as part of a double planet system, and capture after formation elsewhere in the Solar System—are flawed. The group cited a fourth hypothesis, proposed at a conference on the origin of the Moon in October 1984. This hypothesis called for the impact of a body the size of the planet Mars with an accreting Earth, splashing a vast volume of material into orbit around the growing planet. The Moon then accreted from this mass of debris. The working group observed that, although this collision–ejection hypothesis has stimulated new research into the formation of the Moon, it has not supplanted older theories. Preliminary calculations suggest that about half to nearly all the debris from which the Moon formed came from the impact body, the group said. "If these calculations are correct, then the collisonal–ejection hypothesis does not straightforwardly explain the close relation between Earth and Moon, inferred from geochemical data." The group added that the composition of the hypothetical impactor and the chemical effects of the impact would have to be assessed. That presumably would require a geochemical analysis of Mars. "There seems to be an increasing appreciation of the complexity and synergism of processes operating during the formation of the terrestrial planets," the group's report said.

Although the Moon appears to be the Rosetta Stone of terrestrial planet formation, it remains a prime target of scientific exploration in the 21st century. At this writing, no coherent program of either exploration or exploitation has been devised. But it is coming. In fact, it would seem to be inevitable for reasons to be discussed later.

S I X

A MANIFEST DESTINY

I N the fall of 1984, advocates of the lunar initiative the Johnson Space Center, NASA headquarters, and the Los Alamos Scientific Laboratory generated a national symposium on "Lunar Bases and Space Activities of the 21st Century." It was held at the National Academy of Sciences in Washington, D.C., October 29–31. About 300 scientists and engineers attended the sessions, where 135 papers were presented on topics ranging from lunar base logistics to the exploration of Mars. NASA was the nominal sponsor; the National Academy was the host; and the Lunar Science Institute, an adjunct of the Johnson Space Center, arranged program details.

The symposium was the largest and most comprehensive projection of American space objectives since the lunar landings and in effect constituted an agenda for the 21st century. Nothing like it had been presented since the First National Conference on the Peaceful Uses of Space in 1961 at Tulsa, Oklahoma, where the space agency briefed the nation on how President Kennedy's call for a manned lunar landing would be implemented.

The main thrust of the symposium was the return to the Moon and its occupation as a scientific and industrial colony. The keynote of the proceedings was struck by James M. Beggs, then administrator of NASA, who had urged President Reagan to authorize the space station. "One does not have to be an historian of the space age to recognize that a return to the Moon would be a rational extension of our program of expanding human activities in space. We expect that by the year 2000 the space station will be equipped with a supporting infrastructure that will enable us to operate routinely at both Earth and geostationary orbits and between them and eventually at distances as far as the Moon and the inner planets," Beggs said.

Several conference participants cited a revival of interest in lunar and planetary missions in the 1980s, as the shuttle began to fly. Apollo could have led to the establishment of a permanent base on the Moon, but did not, "for reasons that are well documented," Paul D. Lowman, Jr., a geophysicist of the Goddard Space Flight Center, told the conference. He referred to the decision at NASA headquarters to abandon the Moon and to retarget American space objectives to low Earth orbit. The Moon had been put on "hold" during the development of the shuttle, which took a major share of the NASA appropriation in the 1970s. With the shuttle in operation, support for a lunar base was rekindled at NASA centers. By 1984 a return to the Moon had become tied to the construction of a permanent space station.

Lowman summarized the scientific rationale. With the assimilation of the scientific results of Apollo, "our knowledge of the Moon although incomplete is comprehensive enough to say that the environmental feasibility of such a base is now established." The dominant topography and gross composition of the main crustal rocks are well enough known to permit detailed planning for a temporary base supporting a few dozen people, he said. However, the feasibility of a permanent, autonomous colony "is far from established at this time," he added.

With a return to the Moon, America would break out of low Earth orbit and commence manned exploration of the Solar System. This was a major theme of the conference. A secondary theme was

sounded with it. There were allusions to long-term human survival as a rationale for developing space travel. It provided the means of escaping a catastrophic accident or conflict that would destroy the biosphere. This idea was expressed by Wernher von Braun in a comment at the news conference July 20, 1969, following the first lunar landing. "Now we are immortal!" von Braun said. He did not elaborate. In the context of an outlook he had expressed before, it was clear that he meant that now humans had the means of seeking sanctuary beyond the Earth if it were threatened with destruction.

Lowman, too, alluded to this idea. In his symposium paper he noted that evidence from the past has pointed to mass biological extinctions caused by the impacts of planetoid- or asteroid-sized bodies on Earth. "The interval between such impacts may be nonperiodic and if so most life on Earth could be destroyed at any time, as has happened throughout geologic history." He cited an observation by the British philosopher J. D. Bernal that, "if human society or whatever emerges from it is to escape complete destruction by inevitable geological or cosmological cataclysms, some means of escape from Earth must be found. The development of space navigation, however fanciful it may seem at present, is a necessary one for human survival."

At least five great biological extinctions appear in the paleontolological record, according to the work of investigators at the University of California, Berkeley. The most recent, 65 million years ago, at the end of the Cretaceous Period, wiped out the dinosaurs and about half of the genera living then. The investigators accounted for it by the impact of an asteroid 6.2 miles (10 km) in diameter that raised a cloud of dust in the stratosphere. They hypothesized that this effect blocked sunshine and lowered temperature long enough to suppress photosynthesis and cause the collapse of the food chain that supported the giant reptiles and other species.

The authors of this hypothesis were Luis and Walter Alvarez, Frank Asaro, and Helen Michel, who published it in 1980.[1] As evidence for the impact theory, they cited the discovery of concentrations of apparent extraterrestrial iridium, a platinum metal rare on Earth. Concentrations of iridium ranging from 20 to 160 times its average occurrence were found in limestones at the boundary of the Cretaceous–Tertiary Periods from Denmark to New Zealand. This range suggested its deposition by a large, extraterrestrial body.

Although a doomsday scenario may lurk in human consciousness as a motive for developing space travel, it would seem to be too remote a prospect to influence any immediate space policy, compared with the political and military competition that motivated Apollo. The history of the lunar base movement, dating back 40 years, suggests another motivation, one rooted in the American experience of the last two centuries. This is the belief in a manifest destiny that has characterized American expansion since the expedition of Lewis and Clark that opened the West. Having established its frontiers, this space-faring society now perceives its future expansion in terms of a permanent presence in low Earth orbit and an outpost on the Moon and Mars.

PRELIMINARY DESIGN

Lunar base scenarios assume the existence of a transportation infrastructure. In the 1960s it consisted of the Saturn 5 launch vehicle, the Apollo spacecraft, and the Lunar Module, which served as a shuttle between lunar orbit and the surface. Since this system has become extinct, a new one, based on the shuttle, must be developed for the 21st century. As mentioned earlier, it will probably consist of a space station–based Orbital Transfer Vehicle and a reusable Lunar Module. Virtually all lunar base scenarios assume the existence of such a system. Pragmatically, however, the cost of the system is a major factor in the cost of establishing the base.

Science Applications, Inc. had drawn a lunar base design in 1983 that assumed a two-stage Orbital Transfer Vehicle based in low Earth orbit. Each state would be powered by two RL-10 (Centaur) hydrogen–oxygen engines with a thrust of 33,000 pounds each. The logistics of delivering crew and base equipment to the lunar surface had been set forth by the consulting firm in its analysis of space station versus lunar base for the National Science Foundation.

In a paper presented at the Lunar Base symposium, Stephen J. Hoffman and John C. Niehoff of Science Applications, Inc. described a site selection scenario for a permanent lunar base. It required an exploration team of four persons making a 30-day traverse of the lunar surface of examine an area with a radius of about 31 miles. The area would have been selected previously from surface images.

The explorers would cover the area in two roving vehicles with trailers, two persons to a vehicle. The vehicles would be equipped to move soil and expose subsurface strata. The party would carry 5291 pounds of scientific instruments to measure and record seismic, stratigraphic, and other physical data. The data would be analyzed in the trailers.

This site survey, requiring months of preparation, would begin with the launches of the rovers, the trailers, an expendable lander, the Lunar Module, and propellant to low Earth orbit. The launch phase would require 12 shuttle flights. Four Orbital Transfer Vehicle flights would move the base crew and equipment to the Moon. The first two OTV flights would bring the rovers and trailers to be landed by the expendable lander and insert a reusable Lunar Module in lunar orbit. The remaining two flights would deliver the crew and propellant for the Lunar Module and subsequently take the crew back to Earth. The LM would remain in lunar orbit and the base equipment would be left for use by future crews.

An operational lunar base would consist of three main modules, which would be buried for thermal and radiation protection. Rovers left by the site selection teams would be used to bury the modules in a trench. The three main modules would be linked by an airlock-interface module and supplied with power from a 100 kilowatt nuclear reactor. The research crew at this stage would number seven persons: six scientist–technicians and a pilot. They would serve a four-month tour of duty. The pilot and three scientist–technicians would be replaced at intervals.

Two unmanned logistical resupply missions would be flown each year to replace consumables, requiring 18 shuttle flights and eight OTV sorties a year. The initial equipment inventory would include a chemical processing plant, a key device that would enable the scientists to determine how much usable material could be extracted from the soil.

Extensive soil experiments and overland traverses would be carried out with the rovers and trailers. Automated sensing devices, such as seismometers and radiation counters, would be planted at several locations, broadcasting their data to the base and to Earth as well. This method of automated data collection and transmission was successful on Ranger, Surveyor, and Apollo lunar missions and on the Viking missions to Mars.

At the base, life science experiments and observations could be conducted. Food production tests would be done. The site would enable scientists to perform experiments in high-energy physics, attempt to sense gravity waves, and measure the solar wind's intensity and composition.

THE ANTARCTIC ANALOGY

This and other lunar base development schemes followed the system of establishing scientific bases and a logistics main base in Antarctica for the International Geophysical Year (1957–1958). In their symposium paper on "Strategies for a Permanent Lunar Base," Michael B. Duke, Wendell W. Mendell, and Barney B. Roberts of the Johnson Space Center perceived a lunar base as the analogue of the U.S. Naval Air Facility at McMurdo Sound, Antarctica, the American logistics center for the U.S. Antarctic Research Program.

"The lunar base would provide logistical and supporting laboratory capability to a rapidly expanding knowledge of lunar geology, geophysics, environmental science and resource potential through wide ranging field investigations, sampling and placement of instrumentation," they said. "The challenge of long-term self-sufficient operations on the Moon can spur scientific and technological advances in materials science, bioprocessing, physics and chemistry based on lunar materials and reprocessing systems."

In the antarctic, the continental ice cap provided a stable platform from which to view upper atmospheric and space phenomena ranging from ozone concentrations to cosmic rays and magnetic storms —beyond the physical properties of the ice and the land beneath it. It was in the Trans-Antarctic Mountains that Sir Ernest Shackleton and his men toiling toward the polar plateau up the Beardmore Glacier in 1907 found anthracite coal. It was the earliest clue in that region to a discovery that a half century later was to revolutionize geophysics —plate tectonics—for it was hard evidence that in the far past the continent that lay beneath the ice had moved poleward from a temperate zone. Much later, American paleontologists found the fossils of temperate–subtropical, creek-dwelling animals they classed as ancestors of the dinosaurs.

It was not until the development of the ice breaker ship, the airplane, and reliable tracked surface vehicles that long-term antarctic exploration became practical. Lunar exploration similarly requires a transport infrastructure specific to the environment. At antarctic stations, such as Amundsen–Scott at the geographic south pole, persons wintering over are isolated from the mainstream of society for months. Lunar base crews face a similar situation and the psychological effects are expected to be similar. However, at this point, the analogy loses relevance. Although as rich in natural resources as any other continental land, Antarctica has not been developed because even after nearly a half century of exploration the economic prospects are poor, relative to those in other regions. There is reason to suspect that oil reserves are as great there as in Alaska. But the cost of reaching them through marine ice packs or grounded ice a mile thick and of transportation have made the prospect of oil recovery full of risk. As for coal, the difficulties of mining it and moving it across the ice cap make such efforts appear impractical. There is no local market.

If such logistical problems inhibit development in Antarctica, which is fairly well explored and which has been occupied, though sparsely, for three decades, what promise does the Moon, a quarter of a million miles away, hold for economic development? In terms of rationalizing the lunar initiative, the antarctic analogy breaks down. Antarctica and the Moon are not comparable in terms of economic resources, potential, or industrial prospects.

One might begin by saying that one is the last land frontier on Earth and the other, the first land frontier beyond the Earth. The Moon has no visible resources of ice or water, no oil, no coal, no life in or around it. But it does have resources of great potential industrial value shared by no place on Earth. It has low gravity, one sixth of that on Earth, and high vacuum. With these environmental conditions, with its seismic stability and relative nearness to Earth, the Moon offers to provide an industrial annex with a uniquely advantageous environment. Whereas Antarctica is the end of human expansion over the Earth, the Moon is the place where human expansion begins into the Solar System.

A LUNAR POLAR BASE

The location of the initial lunar base has been a matter of speculation for years. It has been assumed generally that an initial base would be located on the front side of the Moon, probably near one of the Apollo landing sites. However, the advantages of a polar base were cited in a symposium paper by James D. Burke of NASA's Jet Propulsion Laboratory.

Because the Moon's polar axis is inclined only 1.5 degrees off the perpendicular to the ecliptic plane, there are no seasons on the Moon. Burke said that parts of the polar regions, especially the crater bottoms, are permanently dark. In those areas where the sun never shines, primordial ice may have existed throughout the ages. Also, in the polar regions, there may be places where part of the solar disk is always above the horizon. In such a region, a solar power plant would provide continuous electricity except during brief eclipses by Earth. If a base were located there, it could rely on solar power and pipe in sunlight to the habitat.

The JPL scientist said that access to a polar base

from lunar orbit would be relatively easy. A landing could be made every two hours from a polar orbit. The general polar landscapes are fairly well known from photo-maps made by *Lunar Orbiter IV* cameras. They obtained a resolution of about 300 feet in sunlit areas. The topography appeared rugged, more like the lunar highlands than the maria basins, Burke said. The permanently shadowed regions of the poles remained invisible.

Because the Apollo missions circled the Moon at near-equatorial (low) latitudes, geochemical scans were not made of the poles. Their chemical abundances would not be determined until the Lunar Geoscience Orbiter is launched, a long-sought and delayed project. Thus, the question of whether ice exists in the shadowed polar areas has been nagging lunar investigators for nearly a generation. It is the only possible source of water and for a while it was thought to be the only possible source of hydrogen. (As mentioned earlier, recent studies show that hydrogen desposited by the solar wind may be recovered from the soil.)

In addition to preserving ice through selenogic time, temperatures as low as 40 degrees Kelvin ($-387.4°F$) are likely at the poles. It would allow the use of cryogenic telescopes in continuous darkness, Burke said. Such an observing environment exists nowhere else on the Moon or on Earth. A preferred location for a cryogenic optical telescope would be the lunar south polar region, from which the center of the galaxy could be observed. Burke said.

For radio telescopes a polar site has no advantage over a site on the far side of the Moon, he added, but it would be advantageous for continuous viewing of the sun. This type of solar observation is available in the antarctic during the austral summer, when the sun never sets, but moves in a slow circle around the polar horizon during the day and night.

Burke projected an image of the polar environment on the Moon. "As the glaring sun creeps endlessly around the horizon," he wrote, "most surfaces are dark unless illuminated by lights or by a mirror. Earth hovers in one direction moving from side to side and up and down a few degrees, but remaining below the horizon from many nearby areas."

THE DESIGN OF A LUNAR BASE

An architectural design study for a permanent lunar base was presented to the symposium by Peter Land, architect and designer of the Illinois Institute of Technology College of Architecture, Planning, and Design. A fundamental factor considered by Land and other planners in the design of a habitat is radiation protection from cosmic rays. Without an atmosphere or magnetic field strong enough to shield the surface, the Moon is constantly awash with solar and galactic cosmic ray particles (atomic nuclei), which may become lethal with long, cumulative exposure. Members of the Apollo crews on lunar voyages reported strange, sudden flashes of light, which were interpreted by NASA physicians as cosmic rays impacting the optic nerve and brain. These were apparently particles energetic enough to have penetrated the metal hull of the Apollo Command Module.

The requirement for radiation shielding of long-term habitats on the Moon imposes a new construction parameter. First-generation structures would consist of pressurized enclosures under canopies of radiation shielding. Shielding would consist of soil and rock fragments of the lunar regolith spread over a supporting structure and raised above the surface. It could be extended at the perimeter as the base expands. When the cosmic ray shield is in place, pressurized enclosures independent of it could be set up underneath. They would consist in part of inflatable tubular beams, light in weight, manufactured on Earth and packed in small volumes for transport to the Moon. On the airless Moon, heavy construction materials are not required. The only breeze that blows across the surface is the solar wind. Part or all of the shielded area could be pressurized as needed for habitats, laboratories, or storage. When the shield is up, Land explained, variations in height can be made by excavating the surface with a drag line.

Outside the pressurized area, instruments and equipment such as heat exchangers, antennas, telescopes, the chemical laboratory, and the power plant could be set up. The Illinois Institute of Technology architect said that this design would minimize the need for heavy construction equipment. The drag line, for example, would consist of a series of scoops pulled across the surface on an endless cable running over one motorized and one free capstan. The architect estimated that the intensity of cosmic radiation at the lunar surface would require 1.2 to 2 meters (3.9 to 6.5 feet) of regolith to reduce radiation exposure to "acceptable limits," which Land defined as acceptable x-ray dosages to terrestrial workers. All lunar operations could be done beneath the shielding, in the pressurized or unpressurized areas. In an unpressurized area, the operator would wear a space suit. In determining the total dosage of radiation occupants of the base would receive during their tours, designers had to take into account secondary neutrons generated by the cosmic ray bombardment. Storm cellars could be provided with extra-thick overburden as safe havens against high-intensity radiation from solar flares. If extra overburden imposed too much stress on the shield, safe havens could be developed in caverns or tunnels, the architect said.

Radiation protection is "pivotal to lunar base design," Land emphasized. The risk should be measured not only by cumulative radiation exposure outside the shelter but also by adding low-level radiation seeping into the shelter, he said. In the Apollo era, cumulative radiation was reckoned for the total time the crew spent outside the Earth's magnetic field, which wards off all but high-energy cosmic radiation. The field extends outward about 40,000 miles from the sunward side of the planet. Reckoning the total time spent outside the field added six days of flight time to the time Apollo astronauts spent on the Moon or in lunar orbit. NASA reported that the total dosages received by lunar flight crews did not exceed levels of industrial radiation deemed safe for individuals.

Shelter design should minimize exposure in order to "maximize the time a person can work outside the shelter," Land said. The base should be consolidated under one shield so that exposure is avoided by persons moving from one part of the base to another. Edges of the shielding must be protected by regolith to prevent horizontal infiltration of radiation. The entrance should be labyrinthine, with overlapping, screened walls.

CONSTRUCTION ON THE MOON

A lunar base construction process adapted to the lunar environment was described by Land in a symposium paper. Support structure bays would have to be erected quickly to provide a radiation-free workplace, he said. This would require that support structure components be fabricated on Earth and brought to the lunar site for installation. The support structure would consist of a floor resting on a lattice of aluminum girders. The regolith would be loaded on the floor, which would then be raised by pneumatic jacks to become the ceiling. Beam ends of the lattice would be connected to columns. The process of raising the floor structure with its load of rocks and soil for radiation protection would be carried out in sections.

The overburden could be leveled after the platforms are raised. This strategy would eliminate lifting operations to place the regolith on the floors after they are raised, a procedure that would require lifting equipment and the expenditure of more energy than raising loaded platforms with pneumatic jacks.

Several types of floor material were considered. One type was moulded regolith, prestressed from end to end with stranded fiberglass tendons by means of small, portable hand jacks. The moulded regolith components would be assembled flat on a leveled ground surface with the tendons inserted and prestressed to form narrow floor sections. These would be connected at the ends to transverse girders which could be manufactured on the Moon.

The architect also considered folded aluminum floors, using all-aluminum components. These would be fabricated on Earth with a profile to permit nesting for transportation. Their length would have to conform to the 60-foot shuttle cargo bay. The prefabricated aluminum floor system would facilitate quick construction of the first section of the shield. Later, with the installation of a production plant, the base team could fabricate floor units moulded from regolith.

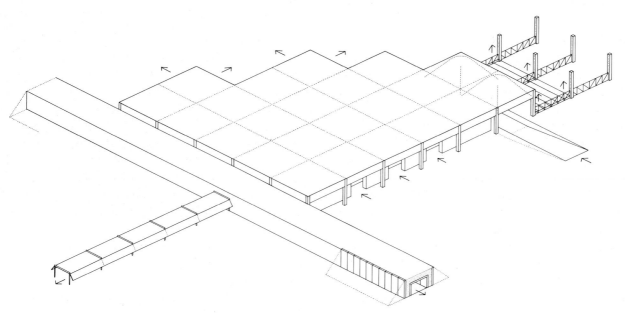

Figure 6.1. A design for a lunar base was presented at the NASA-AIAA Symposium on Lunar Bases and Space Activities of the 21st Century April 5–7, 1988 by Peter Land, Illinois Institute of Technology College of Architecture Planning and Design. In this drawing 1, an erection sequence of a lunar base is shown, right to left. At the right, the sequence shows the columns and aluminum lattice girders in place. The floors are covered by lunar regolith (soil). The floors are then raised by jacks to become the roof, on which the regolith overburden provides cosmic ray shielding. An entrance is provided through overlapping, radiation barrier walls.

Alternatively, inflatable beams could be used for flooring. Land said that pneumatic beams have been developed by Army engineers to bridge gullies and craters in rough terrain for trucks and tanks. These beams consist of large-diameter, long, inflated tubes and smaller cross tubes supporting an aluminum deck. This floor system might be used for the initial section of the shield.

Support for the shield might be a low arch. Land said that it would function in compression without tensile stresses. Reinforcement would not be necessary. The components could be made of moulded regolith assembled over a movable pneumatic support form. Arches would be assembled in sections, each with the width of the form and each embracing several rings of components. Pressurized enclosures would be located underneath the support form.

These enclosures would be interlocking transversely and longitudinally and would self-align under stress. They could be fabricated in a lunar plant from presorted regolith, moulded to a required configuration. The material would be fired to reach a temperature high enough for sintering and surface sealing by means of a solar or electrical furnace or by using epoxy or portland cement to bond regolith aggregates.

The low-arch shield with pneumatic support would support the regolith permanently. The structure would be laid deflated on a level surface and regolith would be pushed over it. Then the structure would be inflated and the elevated regolith evened out. The upper structure surface would be ribbed to anchor it into the regolith. This system could be developed in sections to form a continuous low arch or a single domed structure, Land said.

Plastics vs. Space Station Modules. Land visualized three types of pneumatic structures under shielding canopies. One is supported by air—a structural membrane inflated under pressure. Another is an air-inflated structure with beams, columns, and arches that are independently pressurized to support the membranes between them. A third is a hybrid cable mesh containment system, providing additional membrane support to accommodate higher stresses than other types. Rigid elements could be incorporated in the membranes.

Land observed that recent advances in flexible plastics with high weight-to-strength ratios make pneumatic structures "attractive for lunar habitat and other pressurized uses under radiation shield-

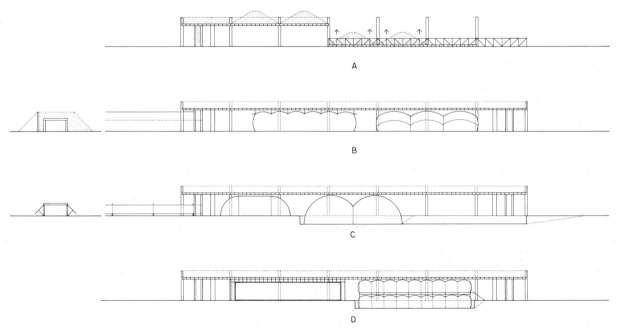

Figure 6.2. In this overview, the shielding is raised in sections. Pressurized enclosures are erected beneath the shielding.

ing." Thousands of small and large pneumatic structures have been erected in many countries since 1950, he said. The construction method Land proposed would eliminate the transport of space station–type modules for shelters by substituting plastic pneumatic structures and shielding platforms. Even though constructed of aluminum, space station modules would require shielding for long-term use, he said.

Land as well as other designers projected the radiation problem as one of the major environmental threats to human habitation on the Moon. This problem was analyzed in detail at the lunar base symposium held by James H. Adams, Jr., of the Naval Research Laboratory, Washington, D.C., and Maurice M. Shapiro, of the Max Planck Institut fur Astrophysik, Munich.

In their paper the authors said that the average intensity of cosmic rays with energies greater than 10 million electron volts may vary 2.5 times from minimum to maximum during the eleven-year solar activity cycle. The composition of galactic cosmic rays, normalized at silicon, is comparable to that of the Solar System in the most abundant elements, they said. However, odd elements generally tend to be overabundant in the stream from the galaxy, because they pass through interstellar gas before reaching the Earth–Moon system.

Occasionally, radiation intensity increases at the Moon because of solar energetic particle events lasting from hours to days. The intensity of radiation at these times ranges from the limits of detection to 70,000 times the intensity of galactic cosmic rays. A solar flare watch will be an important requirement for the safety of any lunar expedition, the authors said, as it was during Apollo.

If a permanent base is established on the Moon, the 5 REM (radiation equivalent man) annual exposure limit for radiation workers on Earth might be an appropriate standard for radiation protection at a lunar base, Adams and Shapiro said. Adhering to this standard, they said, would make it necessary to bury a lunar habitat beneath several meters (six feet or more) of lunar regolith and limit human activity on the surface to 1800 hours a year inside an enclosed vehicle. Outside work would have to be restricted as well. People working on the Moon would have to remain near the habitat, so that they could reach a shelter promptly in case of a solar energetic-particle event (SEP).

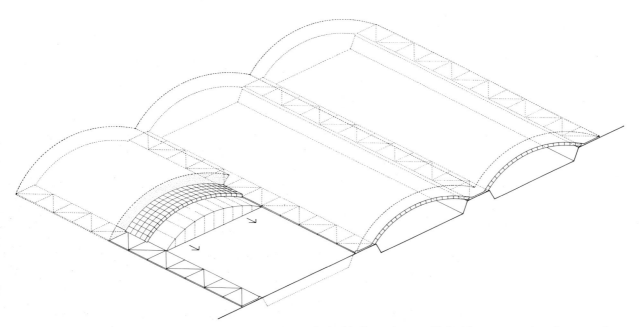

Figure 6.3. An alternate base construction concept is the low arch shield. The arch is established by a pneumatic arch support form. Moulded regolith components are interlocked and laid over the form. Regolith is pushed over the arch and the pneumatic support form is removed. Pressurized enclosures are erected beneath the arch.

At the solar minimum (level of activity), the annual dose equivalent due to cosmic rays is about 30 REM (six times the annual exposure limit for terrestrial workers). Over the 11-year solar activity cycle, the cumulative dose was calculated at 1000 REM, according to Naval Research Laboratory data. Thus, the authors noted, long overland expeditions would be risky unless shelters were provided at intervals of a few hours travel time apart, so that a traverse party could reach one when warned of the sudden onset of a solar flare.

In addition to the hazard to human health, the lunar radiation environment affects electronic systems with potentially serious results. Radiation produces changes in electrical conductivity and causes shifts in device thresholds that may cause a malfunction in an electronic circuit, the authors said. Although electronic components can be built to tolerate large doses over a 10-year lifetime on the Moon, an ionizing particle can change the logic state of digital microcircuitry and damage the information stored in it. This kind of damage is called a "soft upset." If it occurs in a microcircuit of a computer program memory, the program will no longer be operable and must be reloaded, the authors said.

If it is in the microprocessor address register or program counter, the actions taken by the computer will be unpredictable. Soft upsets in control circuitry can result in "unplanned events" such as thruster firings on a spacecraft. A single soft upset could cause the loss of equipment and personnel. It is unlikely that compact circuits can be made immune to them, the authors said.

An Alternative Structure. A proposal for a lunar research outpost and habitat consisting of space station modules shielded by regolith was presented at the symposium by Jan Kaplicky and David Nixon, Future Systems Consultants of Los Angeles. The shielding would consist of a manually deployed superstructure with a layer of regolith on top. This design would shelter six persons and included a pressurized modular habitat, a laboratory and construction workshop, plus an unpressurized pilot oxygen plant.

The authors said that the thin-hulled space station module would not be able to bear a two-meter load of regolith, even in one-sixth lunar gravity. Consequently, the base would need the superstructure. It would be fabricated in kit parts, to be as-

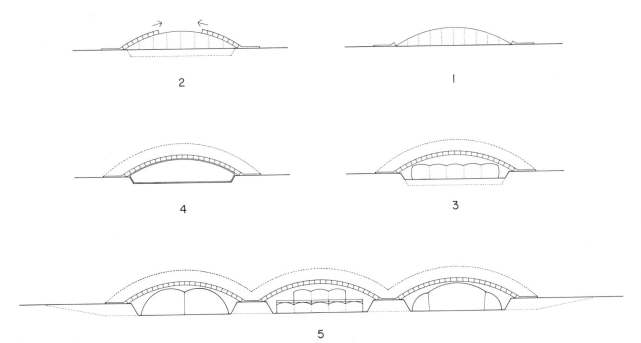

Figure 6.4. Arches can be interconnected to shield a range of pressurized enclosures forming the base.

sembled on the lunar surface by mission specialists. It would be constructed as a rectangular superstructure over six space station–derived modules, each 10 m long (32.8 ft) and 4.5 m (14.7 ft) in diameter. The authors described the envelope as a shallow, flat-topped mound of loose regolith supported by a continuous tension membrane connected to a grid of telescopic columns and tapered beams underneath. The ends would be left open, with protective overhangs.

The arrangement of beams and columns would take the form of an orthogonal grid of structural beams, each 5.2 meters (17 feet) long with a 5 meter (16.4 feet) beam span straddling the girth of the modules. The height would also be 5 meters. A series of high-tensile, fine-mesh membranes would be stretched between the beams to provide support for the regolith mass above. The mesh would be made of woven graphite fiber sized to allow a deposit of 1-mm-grain-sized regolith and larger sizes. The regolith would be deposited over the envelope by a remotely controlled, mobile conveyor system, delivering loosely compacted material, bay by bay.

Vertical column and angled beams would be made of advanced composites, derived from graphite epoxy or similar technology, developed for the space station's deployable truss structure. Each column would be connected to a circular footpad to spread the load. The columns would be designed as telescopic tubes to facilitate low, preassembly of all beams to columns at node points and the attachment of mesh membranes to the beam edges.

The main beams would span 5 meters laterally (side to side) and 2 meter struts (6.5 feet) would support the frame longitudinally by being placed at right angles to the beams at 2.5 meter intervals (8.2 feet). Cross bracing would provide longitudinal stiffening.

The plan calculated that an envelope 546 meters square (5870.97 square feet) would be required to shield the six modules and equipment. It would be a rectangle 26 by 21 meters (85.3 by 68.8 feet). Two rows of three modules each could be emplaced side by side beneath it, pointing toward the open ends of the envelope.

Hardware required for the shelter included 56 telescopic columns, 84 footpads, 70 main beams,

143 struts and 65 woven mesh panels. Before being shipped to the Moon, the structure would be tested in a terrestrial (1-G) environment.

The authors calculated that the complete structure would have a launch weight of 7500 kilograms (16,500 pounds) and could be stowed in a cylinder 5.5 meters (18 feet) long and 2.5 meters (8.2 feet) in diameter. It could be delivered to the Moon on a single mission, they said.

A tractor would be needed to move the base modules from the landing site to the base site. A mechanical conveyor would be required to deposit the regolith on the upper surface of the canopy. The tractor and conveyor would be delivered on a separate mission. The superstructure would be set up by mission specialists. Beams, struts, columns, and footpads would first be assembled at shoulder level and mesh membrane panels would be attached. Each structural bay would then be raised to the planned height by manually operated screws with crank handles. After the superstructure was leveled and anchored, the regolith would be deposited on top. The modules would then be moved into position beneath the canopy by the tractors.

The Concrete Scenario. The application of concrete made of lunar soil for the construction of a lunar base was presented by T. D. Lin, Construction Technology Laboratories, Portland Cement Association, Skokie, Illinois. In an introduction the author explained that cement basically consists of limestone, clay, and iron ore. Limestone, a deposit of marine life, does not exist on the Moon; consequently, another source of calcium silicate would have to be used. It could be obtained by processing the lunar regolith.

Most concrete, Lin said, is made with Portland cement, consisting typically of 25 percent calcium oxide, 23 percent silica, 4 percent alumina, and a small percentage of other inorganic compounds. The author noted that data from Apollo lunar soil samples showed that they had sufficient silica, alumina, and calcium oxide for the production of cementitious material. Calcium oxide could be obtained from anorthosite, a calcium-rich feldspar containing about 19 percent calcium oxide. Lin referred to Lunar Sample 60015, a shock-melted anorthosite rock coated with glass, as the type of material that could be used for cement. The glass is a potential cementitious material if ground to fine particles.

Concrete is an amalgam of cement and aggregates. To produce it on the Moon, lunar rocks would be crushed to suitable, coarse, aggregate size and sieved to the gradation of fine aggregates. All the ingredients are present except water, which is required to make the cement paste that binds the aggregates into a rocklike mass as it hardens. Oxygen, the principal part of water, is plentiful on the Moon, bound up in rocks, but hydrogen would have to be imported from Earth unless enough could be collected from solar wind deposits or unless ice is found at the lunar poles.

Lin suggested that lunar ilmenite, a mineral composed of iron, titanium, and oxygen, could be heated with hydrogen imported from Earth to release steam as a source of water. The residual iron could be processed into fibers, wires, and bars to reinforce the concrete. Further, Lin said, casting and curing chambers for concrete could be made from empty space shuttle external tanks—if they were lifted to the Moon instead of being dumped into the Indian Ocean after every shuttle launch.

In making the case for concrete shelters on the Moon, Lin said that while the flexural strength of concrete is low (compared with its compressive strength), flexural strength can be increased by reinforcing the mix with steel or glass fibers. The fibers prevent cracking. Concrete reinforced with 4 percent (by weight) of steel fibers has nearly twice the flexural strength of plain concrete, Lin said. On the Moon, the author added, lower gravity and the lack of wind and moonquakes would allow a greater length in the span of a flexural member.

Lin illustrated the shape of concrete structure might take on the Moon with a sketch of a three story building 210 feet in diameter. Each story would be 15 feet high. The structure would be strong enough to contain one atmosphere pressure inside against vacuum outside. A cylindrical tank at the center would serve as safety shelter in case of damage by meteorite impact or some accident causing a leak. The tank would also provide a refuge from high-energy cosmic ray particles during a solar flare.

The roof of the building would be shielded by 6 to 18 feet of regolith, to protect occupants and equipment from cosmic ray effects.

According to the author's calculations, the major stress on the building would be internal air pressure against the walls. The pressure can be countered by using circular panels facing outward and supported by columns. That configuration would change tension into compression, to which concrete is highly resistant. Steel tendons would secure the columns into position. The tendons could be wrapped around the cylindrical tank at the center of the building and stressed to provide hoop forces on the tank. They could then be anchored to columns at the opposite side. At the wall faces, panels would be installed to contain the soil for insulation against radiation and the large temperature changes of the long lunar days and nights. These range from minus 292 to plus 212 degrees Fahrenheit.

The concrete "house" in Lin's scenario has 90,000 square feet of usable area. It would require 250 tons of steel and 12,200 tons of concrete, consisting of 1500 tons of cement and 490 tons of water. All the material is readily available on the Moon except 55 tons of hydrogen (for water), which may be recovered from solar wind deposits.

AN ADOBE VILLAGE

Another lunar shelter concept offered adobe construction. It was presented in a paper entitled "Magma, Ceramic and Fused Adobe Structures Generated in Situ" by E. Nader Khalile of the Southern California Institute of Architecture, Santa Monica. The author suggested that lunar structures could be cast directly from mounds of regolith by melting the regolith with focused sunlight and allowing the magma to be cast into a shell. The underlying loose soil in the mound would then be excavated and packed over the shell for radiation, thermal, and impact shielding. Khalile referred to earlier papers which had described such a process.

The heat would provide a semiglazed interior that would be airtight, the author said. Conventional structures could be built by fabricating structural beams, columns, panels, and connections from magma, reinforced by regolith-derived fibers or mesh. Khalile cited the space shuttle's heat shield tiles as an example of how ceramic material can be used for lunar applications. Ceramic structures of limited span could be cast on lunar sites or generated in space. On the Moon, a centrifugal gyrating platform—a giant potter's wheel—with adjustable rims and high flanges could be used for dynamic casting of ceramic and stoneware structures, the author said. In space the "potter's wheel" would create more varied ceramic forms than in a 1-g field. A mass of regolith could be dumped into the stationary center zone of the platform and melted by focused sunlight to flow to the periphery rotating zone and cast into desired shapes. Known lunar material could be spun on the same platform to create fiber. By integrating the two operations, monolithic ceramic structures with tensile fiber reinforcing layers could be generated.

Double-shell ceramic structures sandwiched with insulating material (or space) could provide radiation, thermal and impact shielding. The gyrating platform, the author suggested, could generate ceramic modules in situ. The raw material is soil—whether on the Moon, an asteroid, or the planet Mars.

In situ structures could be made of adobe blocks produced from unprocessed regolith or the tailings of mining operations, the author said. The blocks would be formed by fusing regolith with solar heat. The low gravity field and high vacuum of the Moon would allow a smaller angle of repose and enhance cohesion.[2] Fusion of the top layers of soil could form a lava crust for landing pads and roads.

Cave Dwellings. A proposal that lava tubes or caves left by the outflow of molten rock from primordial melting of the lunar surface can be developed as lunar shelters was presented by Friedrich Horz, Johnson Space Center. Lava tubes, known in terrestrial volcanic regions, may be extensive on the Moon. This has been indicated by surface photos showing long, sinuous rilles, a magma formation similar to lava tubes but open at the top.

Horz suggested that lava tube cross sections could be modified and enlarged by heating to provide a

Figure 6.5. A surface exploration crew begins its investigation of a typical, small, lava tunnel to determine if it could serve as a natural shelter for lunar base modules. Tunnels such as this are left when the surrounding lava cools, forming the walls and roof. Artwork by John Lowery/Courtesy Lockheed/NASA.

smooth surface for walls, floor, and ceiling. If rilles are an indication, lava tubes on the Moon may be tens of kilometers long and hundreds of meters wide and deep, he said. However, the only lava tubes that can be seen in photos are those with partially collapsed roofs. But judging from crater observations, some lava tube roofs are several tens of meters thick. "There is little doubt that lunar lava tubes have enough cross section to house almost any habitat," said Horz.

If natural caverns large enough to house an entire lunar base exist on the Moon—and Horz said it appeared to him that they do—they would provide safe and long-term shelters against radiation and meteorite impact. Roof thickness might be more than 10 meters (32.8 feet), he said. Caverns or tubes would provide shelter for pressurized modular habitats brought from Earth or fabricated from lunar regolith. Inflatable habitats also could be used, he said. The tubes would provide a relatively constant temperature at minus 20 C (−4 F), in contrast to the lunar day–night temperature cycle with a range of 500 degrees F.

Oxygen Production. The production of oxygen from Moon rocks was described in a symposium paper on "Environmental Considerations and Waste Planning on the Lunar Surface" by Randall Briggs and Albert Sacco, Jr., Worcester Polytechnic Institute, Shrewsbury, Massachusetts.

They based their oxygen plant at a lunar station with a crew of 15 whose main function was extracting oxygen from lunar ilmenite ($FeTiO_3$) by a hydrogen reduction process. The source of power was an SP-100 nuclear reactor generating 100 to 400 kilowatts. The crew would subsist in a semiclosed life support system, supplied with food from Earth.

The authors selected ilmenite, an iron–titanium-rich mineral, because of surveys during the Apollo program that it constitutes about 10 percent by weight of the dark-hued basaltic maria surfaces. Their calculations indicated that oxygen production on the Moon could become big business, supplying the needs of NASA not only on the Moon but also in low Earth orbit and geostationary orbit.

They estimated that for every 200,000 kilograms

of liquid oxygen (lox) extracted from the basalt, about 54,000 kilograms net would reach low Earth orbit after the balance was used in transporting it. On the basis of this ratio of 3.7 to 1, it would take annual production of 2.1 million kilograms of lox to provide NASA's annual requirement of 552,000 kilograms in low Earth orbit for lunar and geostationary orbit transportation. The NASA lox requirement would exist at the time the lunar production plant was in full operation—some time in the early 21st century.

The reduction of ilmenite with hydrogen involved a single chemical equation: $FeTiO_3 + H_2 \rightarrow Fe + TiO_2 + H_2O$ (iron–titanium oxide and water). The water vapor is electrolyzed to hydrogen and oxygen and the hydrogen is recycled and reused in the process. Producing 2.1 million kilograms of lox requires 20,000 tons of ilmenite from 110,000 cubic meters of bulk soil a year.

The authors calculated that a semiclosed life support system for the 15-person station complement would consume 13.5 kilograms of metabolic oxygen a day, 54 kilograms of drinking water; 81 kilograms of hygienic water, and 9 kilograms of food. The daily waste was calculated at 15 kilograms of carbon dioxide, 37.4 kilograms of water vapor, 22.5 kilograms of urine, 2.4 kilograms of feces, and 189,900 kilojoules of metabolic heat.[3]

The life support system would use oxidation water reclamation. The authors cited a 1975 study (Robert Jagow, Lockheed) showing that wash water, human waste, and trash can be oxidized at high temperatures and pressures to produce water, carbon dioxide, and a cake of phosphates and sodium salts. The study indicated that the water may be recycled for human consumption. The carbon dioxide could be reduced to oxygen and carbon, producing 1500 kilograms of pure carbon a year.

Subselene. Another approach to base construction was set forth in a symposium paper by John C. Rowley and Joseph W. Neudecker of the Los Alamos National Laboratory. It was a rock- and soil-melting process of excavation with an electrically heated tunneling machine. The machine was designed to bore through rock by melting it, leaving the tunnel lined with impermeable glass.

This process is based on a research and development project for tunneling developed at Los Alamos and demonstrated successfully for a number of years. The authors suggested that the technology called Subterrene could be applied to lunar base construction, drilling, and coring on the Moon (as Subselene). They cited these features: The melting method is insensitive to various types of rock or soil conditions and can be automated for remote operation. Electrical or nuclear energy can be used for resistance heaters on penetration devices.

Terrestrial basalts similar to lunar basalt melt at 1570 degrees Kelvin (2366.6 F). With one type of penetrator, the authors said, all rock melted during hole formation would be densified to form a glass lining and no debris removal would be needed. Another type of penetrator, called an "extruder," would have ports, allowing the melted rock to flow back through the head into a device that would chill the melt and form debris cuttings. These solids could then be formed as glass pellets, rods, or wool. Tunneling would be more efficient for building subsurface structures than trenching or back covering, the authors said. They described one construction method using a large tunneling machine with a ring melting penetrator, powered by a nuclear reactor. Once the tunnel was made, bulkheads of lunar glass would provide sections for habitation, storage, and laboratories. Another method excavated a mound of lunar regolith under a glass roof formed by a portable Subselene machine to create a room with an arch and walls of melted rock. A prototype of such a structure has been demonstrated at a terrestrial site, the authors said.

The processing of lunar materials by microwaves was described in a symposium paper by Thomas T. Meek, David T. Vaniman, Franklin R. Cocks, and Robin A. Wright.[4] They explained that extra-high-frequency microwaves between 100 and 500 gigaherz have the potential for extracting specific materials, such as oxygen, from the soil. They can be applied for melting selectively and used in preparation of ceramics, like bricks, or for the direct preparation of hermetic walls in underground structures. They added that microwave processing could be used to produce photovoltaic devices from lunar material, especially from ilmenite. Preliminary experiments in ultra-high-frequency melting of an ilmenite-rich terrestrial rock shows that microwave processing is feasible in these rocks, they said.

Ultra-high-frequency microwaves at 2.45 gigaherz and extra-high-frequency microwaves at 100 and 500 GHz would have advantages over focused sunlight and other heating methods, the authors said. Besides saving energy, they would make possible selective heating of desired rock phases. This offers the prospect of designing continuous flow from the raw material to the finished product in a self-contained process.

Large rocks could be fractured by coupling the energy to rock phases with large thermal expansion coefficients. Then the fractured rocks could be melted and separated into other raw materials or used directly for fabrication into simple but useful shapes, such as bricks. With sufficiently high temperatures, it is possible to decompose lunar material into its constituent elements without the need for any chemical feedstocks or further electrochemical processing, the authors said.

The authors noted further that UHF microwaves at 2.45 GHz couple strongly to water. If moisture exists at the lunar poles, it could be recovered by microwave heating of rocks if it were absorbed in the rocks and was not actual ice. Inasmuch as oxygen is 40 percent by weight of lunar soil, the extraction of oxygen by microwave heating would be of "great utility" in the long-term support of a lunar base.

Hydrogen, too, might be recovered with microwave energy, from solar wind deposits, the authors added. If the average abundance of hydrogen is 50 milligrams per gram of surface soil, an efficient microwave extraction system could obtain 50 grams of hydrogen from a metric ton of soil, to be converted to 0.45 liters of water.

CELSS A controlled ecological life support system (CELSS) was discussed in a symposium paper by R. D. MacElroy and Harold P. Klein, Ames Research Center, and M. M. Averner, University of New Hampshire. The system was based on the regenerative use of oxygen, water, and carbon

Figure 6.6. NASA has built this Biomass Production Chamber in Hangar L on Cape Canaveral as an experiment in hydroponic food production in a space station, space ship, lunar or extraterrestrial planetary environment. The Controlled Ecological Life Support System is designed to grow enough food to sustain one adult person indefinitely. Eventually, such a module will be flown in the space station laboratory. (NASA)

Figure 6.7. This interior view of the Biomass Chamber shows a wheat crop almost ready for harvest. The plant growth experiment began in 1987. (NASA)

Figure 6.8. Dr. Walter Hill, a plant scientist of the Tuskegee University, Alabama inspects sweet potato growth in the soil-less, hydroponic "farm" in Hangar L. (NASA)

Figure 6.9. Early results of the Biomass Production experiment are shown in these trays. Wheat in the center tray is 42 days old; in the right tray, 33 days.

dioxide. As mentioned earlier in this chapter, carbon dioxide produced by the crew could be concentrated and processed to release oxygen. Used water in urine and wash water could be reclaimed by removing dissolved or suspended material. Equipment for these processes has been developed at two NASA centers, Ames and Johnson.

A CELSS module would supply the crew with water, food, and oxygen. It would be capable of growing algae and food plants that consume carbon dioxide and water and convert them into food with light. Microbes would help reduce nonedible plant fractions (or garbage) in an ecological system. Genetic engineering could increase productivity.

The authors said that the first lunar outpost could have a bioregenerative life support module. They expected that such a module would be added to the space station.

The 1984 picture of a bioregenerative life support system for a lunar base was a small, highly automated one capable of intensive agriculture. The optimum food crop for a small "farm" of 10 square meters (107.5 square feet) is wheat, according to the authors.

In a paper on "Wheat Farming for a Lunar Base," Frank B. Salisbury and Bruce G. Bugbee, of Utah State University, reported that their studies of wheat production showed that minimum growing area per person would be about 6 square meters (64.5 square feet), using 3.55 kilowatts of electrical energy. A lunar farm the size of a football field could support 100 people on that basis.

Since 1986 a Controlled Ecological Life Support System has been developed as a Kennedy Space Center project in Hangar L, a former storage building, on Cape Canaveral. The project scientist, Dr. William M. Knott, plant physiologist, and a team of agricultural engineers and technicians have grown crops of soybeans, wheat, and white and sweet potatoes hydroponically in their Biomass Production Chamber at one end of the huge hangar.

The chamber, which is sealed in operation, was adapted from the Spacelab module, the European Space Agency's laboratory unit, that has been flown in the cargo bay of the shuttle orbiter. It is 24.6 feet long and 11.48 feet high with its air and water regeneration and production components. The sealed plant growth chamber has an area of 282.56 square feet. The soybeans, wheat, and potatoes are grown from a porous membrane in plastic or steel tubes through which water and fluid nutrient pass. Holes along the length of the tubes allow the plant stems to grow out of the tube as they seek the light. The plant roots absorb moisture and nutrient through the membrane's 2-micron sized pores. The roots tend to grow around the tube. The plant seed is embedded on the membrane.

The stem grows toward a light "gradient" and the roots move toward a gravitational gradient. Inasmuch as the Kennedy Space Center program is carried on in 1 g, the gravitational force required to attract the roots appears to be speculative. Director Knott said it had been calculated at about .001 g.

As far as NASA's experience has gone, its microgravity tests on plant behavior has been limited to 10 days (STS 9, *Columbia*, November 28–December 8, 1983). The testing sought to find out in part whether plant rhythms of growth and development were exogenous or endogenous. Some decrease in lignin formation was noted in microgravity, Knott said.

"We can't tell yet whether plants will grow and produce food in microgravity," Knot said. "We need at least 90 days in orbit for that." If the Soviets have the answer, NASA researchers are not aware of it. The Soviets have experimented with light levels and nutrients.

Except as an experiment, growing food in orbit hardly makes a contribution to a space station in low or geostationary Earth orbit. Food deliveries would be part of the routine, and water can be readily recycled. However, the capability of growing food on a spaceship becomes vital necessity on an interplanetary voyage to replenish fresh vegetables over years of travel.

Along with the Kennedy Space Center, the Ames Research Center at Mountain View, California, is working on a space station biomass production system that would be adaptable to interplanetary voyaging. In addition to wheat, soybeans, and potatoes, the hydroponic growth crops include lettuce, beets, and tomatoes.

At Hangar L, the Kennedy Space Center's Bio-

Figure 6.10. Soybeans were photographed after 75 days. Plant growth rates and nutrient composition are closely monitored. All light is artificial. (NASA)

Figure 6.11. These growth tubes are designed to allow the plant stem to grow outward toward the light source. The roots entwine about a porous membrane inside the tube. (NASA)

mass Production Chamber grew its first plants in December 1987. Within a year, it had produced an abundance of wheat, soybeans, and potatoes. "This chamber is our first attempt to grow food to sustain one person indefinitely," Knott said. "The chamber can be used for all crops—wheat, soybeans, potatoes, lettuce, beets, even sugar beets, and, of course, tomatoes. They can be grown in combination."

On the space station, artificial lighting would be required in low Earth orbit, where the station would be in sunlight only 45 minutes of every 90 to 96 minute orbit. The effect of intermittent sunshine has not been determined.

The lunar environment presents a radiation hazard. Ultraviolet radiation might be filtered out by a specialized greenhouse. The effect of other types of radiation is not known. It might be simpler to grow plants on the Moon by directing sunlight through tubes or light pipes into subsurface chambers, or "farms." Whether terrestrial plants would grow in lunar soil has not been determined, Knott said. Analysis of the soil returned by lunar landing missions showed that it was sterile.

The current approach of the biomass project is to develop a hydroponic system in a sealed environment, Knott explained. In just two years, it has worked well enough to justify its continuation. The proper chemical and nutrient balance of nutrient fluid is maintained by a computer. "The system need not be buffered," Knott said. "If the pH (hydrogen ion concentration) changes, the plant immediately sees it. The challenge is to the computer to manage the nutrient. You've got to cycle the carbon, hydrogen and oxygen very fast; also, determine the mass and energy flux of these systems."

For the space station, a separated plant growth module would be needed, Knott said, even though it would be carried in the Laboratory module. Waste and food processing techniques and hardware eventually will be integrated with the biomass production unit.

Waste processing, Knott said, will start with control of waste contaminant gas and the recycling of condensate water. Later, the processing of solid and liquid human wastes will be incorporated in the system. Methods by which the inedible portion

Figure 6.12. The wheat plants approach maturity in the chamber. (NASA)

of the plants are converted to edible material will be developed later.

When the controlled ecological life support system becomes fully operational, Knott said, and waste and food processing modules are integrated with biomass production. "we will begin to fully understand the mass and energy requirements of a CELSS." The next step will be a human-occupied chamber for a ground test of an operational, closed, bioregenerative life support system, he said.

BUDGET

The Budgetary Feasibility of a Lunar Base was analyzed by Wallace O. Sellers, Merrill Lynch & Company, and Paul W. Keaton, Los Alamos National Laboratory. They estimated that a permanent lunar base could be established over 20 years for "well under $100 billion in current (1984) dollars." They said that Apollo was carried out in half that time for $80 billion during a period when the Gross National Product was one-half of the GNP (in 1984).* "We concluded that a lunar base program can be carried out in an evolving space program without extraordinary commitments such as occurred in 1961," they said. The basic cost would have to be borne by government.

The Sellers–Keaton study cited a Johnson Space Center forecast that a space transportation system capable of delivering payloads to lunar orbit would exist in the later part of this century. However, transportation equipment developed for lunar exploration would be charged to the base program, the JSC study said. The study estimated that maximum yearly expenditures for hardware and transport would amount to about $6 to $9 billion (1984). Heaviest funding would be due in the period 2006–2010 or later. Sellers and Keaton concluded, "The commitment should be made now; in the long run, it is probably inevitable. The sooner it is made, the more intelligently and economically the lunar base can be planned and implemented. We believe that the program is affordable."

*One of several estimates of the cost of Apollo, 1961–1972, in 1984 dollars.

Figure 6.13. These sweet potatoes and white potatoes, grown without soil, are products of Hangar L farm on Cape Canaveral. (NASA)

Figure 6.14. A Bionetics Corp. Horticultural technician, Lisa Siegriest, and scientist Tom Deschel harvest a tray of dwarf wheat. The crop was grown in the summer of 1989 in the Controlled Ecological Life Support System biomass chamber at the Kennedy Space Center, Florida. Plant growth was computer controlled. The experiment provided further evidence that crops can be grown under controlled environmental conditions in space. (NASA).

SEVEN

POWER FROM
THE MOON

THE two most persistent questions about the Moon concern its origin and utility: Where did it come from? What is it good for? The first question has yet to be answered unequivocally. The second raises another: What does the Moon have that Earth needs? Lunar low gravity and high vacuum are assets for certain types of industrial processes. Otherwise, natural resources on the Moon are also available on Earth, with a notable exception. It is the isotope helium 3, cited earlier by Geoffrey Briggs, director of the Solar System Exploration Division at NASA headquarters, as an important discovery on the Moon.

Helium 3 is known as a potential source of clean energy when burned as fuel with deuterium (in lieu of tritium) in a thermonuclear fusion reaction. The isotope is rare on Earth, where the supply is measured in hundreds of kilograms,[1] but it appears to be abundant on the Moon, where it has been accumulating for billions of years as fallout from the solar wind.

Analyses of space probe data since 1959 have shown that along with many other elements, the solar wind carries a flux of helium particles with a high ratio of the isotope helium 3 to normal helium 4 compared to the cosmic abundance. Because it contains electrically charged particles, the solar wind blowing outward from the corona of the sun is deflected around the Earth by the Earth's magnetic field. But the wind impacts directly the airless surface of the Moon, which has no magnetic field.

There the solar wind has deposited vast quantities of helium 3 along with hydrogen and other elements in the lunar topsoil, according to estimates by scientists at the Fusion Technology Institute of the University of Wisconsin. A paper published in 1986 based on analyses of lunar samples returned by the Apollo expeditions and Soviet Luna collectors projected the current existence of one million metric tons of helium 3 in the lunar regolith.[2]

Institute scientists refer to the helium 3 resource as "astrofuel." They assert that once the technology of fusion power is developed, astrofuel can revolutionize the power economy on Earth and provide a new and more powerful means of space vehicle propulsion than chemical propellants.

The fusion of light elements into heavier ones is the process that releases energy in the sun. Scientists have been trying to reproduce it in the laboratory since 1951. After the investment of more than $20 billion in fusion experiments and reactor designs, the worldwide fusion community is within a few years of the first "break-even" demonstration, according to Gerald L. Kulcinski, director of the university's Fusion Technology Institute at Madison.[3] Early in the 1990s, he said, magnetically confined plasmas in devices at Princeton or at Culham in the United Kingdom are expected to release more thermonuclear energy than needed to start the reaction. As of April 1988, he perceived the situation in fusion research as analogous to the fission experiment by Enrico Fermi at Stagg Field, the University of Chicago, in 1941. That work opened the atomic age.

The present worldwide effort in fusion is concentrated on the reaction of deuterium and tritium, isotopes of hydrogen, wherein, said Kulcinski, 80 percent of the energy is released in the form of neutrons. The drawback of this process is that the neutrons induce radioactivity in structures close to

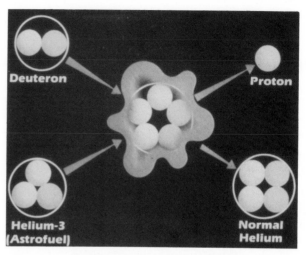

Figure 7.1. The Astrofuel fusion reaction with helium-3 releases 18.4 MeV per event. (College of Engineering, University of Wisconsin, Madison)

the plasma and generate radioactive waste. Kulcinski and his colleagues contend that a cleaner and more efficient reaction is obtained by substituting helium 3 for tritium. A deuterium–helium 3 fusion reaction does not involve neutrons or radioactive species directly, they said, although side-reactions (deuterium–deuterium) release approximately 1 percent of the energy in neutrons. Thus, 99 percent of the energy from the deuterium–helium 3 process can be released as electrically charged particles which can be converted directly into electricity by electrostatic means, they said. Alternatively, the particles can be funneled by magnetic means through a rocket nozzle to produce thrust for a space vehicle.

Compared to fusion with tritium, the low neutron release of the deuterium–helium 3 process "greatly simplifies the safety-related design features of the reactor and reduces the levels of induced radioactivity such that extensive radioactive waste facilities are not required," Institute scientists said. They claimed that the efficiencies of deuterium–helium 3 fusion would be 60–70 percent, compared to conventional thermal power systems with only 30–35 percent efficiency. From that viewpoint the deuterium–helium 3 process is clearly advantageous—except for the cost of extracting helium 3 from the Moon.* However, even at an estimated procurement cost of $1 billion a ton, the energy density and efficiency of astrofuel yield an energy cost equivalent to that of oil at $7 a barrel, the University of Wisconsin's College of Engineering estimated.[4] Inasmuch as deuterium–helium 3 fusion promises electric power production at 70 percent efficiency, the Wisconsin group suggests that it could reduce the cost of electric power in the 21st century. Institute scientists estimated that fusion with helium 3 can produce electric power with twice the efficiency that can be attained in fossil or fission fuel power plants.[5]

In contrast to the direct conversion of fusion energy to electric power, the fossil and fission plants must produce steam to run turbine generators.

* Deuterium is abundant in the oceans and tritium is bred by allowing neutrons to react with lithium-bearing compounds.

GOLD IN THEM THAR HILLS

How much energy is contained in lunar helium 3? Fusion Institute scientists calculated that in 4.6 billion years, the solar wind has dumped more than 250 million metric tons of the isotope on the Moon. Analysis of the lunar samples picked up by the U.S. Apollo expeditions and the Soviet Luna samplers indicate that a million metric tons are still trapped there.

The characteristic of lunar soil that makes it an effective helium collector is its fine grain size, the product of billions of years of meteorite bombardment.[6] Solar wind particles have been implanted in the soil granules. Helium 3, hydrogen and other gases were found in the 43 pounds of lunar soil and rocks collected by Neil Armstrong and Edwin E. (Buzz) Aldrin, Jr., the first men on the Moon, in 1969 and returned to Earth aboard *Apollo 11*. Similar deposits were found on the later missions. Higher concentrations of solar wind particles were found in the mineral ilmenite ($FeTiO_3$), which comprises about 10 percent of the lunar maria and gives them their dark color. The particle concentrations were lower in the light-colored lunar highlands.[7]

Fusion Institute Scientists characterized the recovery of helium from the lunar soil as a simple process, although many tons of soil would have to be mined. Mining sites would be selected with an assessed isotope concentration of 36 parts per million to a depth of two meters (6.52 ft). The soil would be moved by automated excavators on conveyor belts and delivered to a processing facility, where it would be beneficiated (screened to remove large particles for better gas concentrations) by electrostatic and electromagnetic methods and delivered to vacuum degassing equipment. Statistical analysis has shown that about 80 percent of the helium is contained in particles in the 20 micron size range. The solar wind gases would be released by heating.

The production of 1 kilogram (2.2 pounds) of helium 3 would require the mining of 120,000 metric tons of soil. In terms of energy, the yield from the deuterium–helium 3 fusion reaction was estimated at 250 times the cost of energy to mine the isotope on the Moon, process it, and ship it to Earth.

In this calculation an unmanned Earth–Moon transport the size of the space shuttle was assumed to depart from Earth orbit, land on the Moon, and return with a payload of 30 metric tons of liquid helium 3.[8] Compared to the 250-fold energy payback on lunar helium 3, the payback for uranium 235 production is about 20-fold and that for coal is about 16-fold, the scientists noted.

Obviously, the mining of helium 3 would produce other solar wind elements of industrial and commercial value. Institute scientists estimated that for every metric ton of helium 3 extracted from the regolith, 3300 tons of helium 4 would be produced, along with 500 tons of nitrogen, more than 4000 tons of carbon monoxide and carbon dioxide, and 6100 tons of hydrogen. The hydrogen with oxygen would provide water for life support, electrical energy from fuel cell batteries, and rocket propellant. Oxygen is known to be abundant in the lunar rocks. The nitrogen could be used to grow plants in pressurized greenhouses; the carbon, also available from the soil, would be used in manufacturing, and the helium 4 would be used as a power plant working fluid and for pressurization.

With helium 3, the Wisconsin scientists' analysis of the Moon's economic potential is quite optimistic. The million tons of helium 3 is reckoned by the Institute investigators as having a value on Earth of at least $1 billion a ton. Its energy potential is 10 times that contained in all the known economically recoverable fossil fuels on Earth—coal, oil, and gas; 100 times that in the economically recoverable uranium burned in light water reactors; and roughly twice that in uranium used in liquid metal fast breeder reactors. The researchers estimated that 25 metric tons of helium 3 burned with deuterium would have provided all the electricity used in the United States in 1986. This volume, it was noted, would fit into the cargo bay of a space transport the size of the shuttle.[9]

THE MINING SCENARIO

In assessing the prospect of mining helium 3 on the Moon, Fusion Technology Institute scientists reported that 120,000 tons of soil must be mined and beneficiated (screened for maximum helium 3 content) to produce one kilogram of helium 3. They calculated that the regolith contains 1.12 kilograms of the isotope per 40,000 tons, or a ratio of 180 grams per ton. The extraction of the helium would require mining equipment with a mass of 965 tons to be flown to the Moon. It could mine 737 tons of soil an hour, using 1981 kilowatts of electric power, obtained from the sun by solar power generators.[10]

A detailed helium 3 lunar mining scenario was drafted by research scientists Igor N. Sviatoslavsky of the Fusion Technology Institute and the Wisconsin Center for Space Automation and Robotics and Mark Jacobs of the Astronautics Corporation of America, Madison. They presented it in a paper at the Space 88 Conference in August 1988 for the American Society of Civil Engineers, the American Institute of Aeronautics and Astronautics, and associated groups held at Albuquerque, New Mexico.[11]

The Sviatoslavsky and Jacobs concept described an automated mobile lunar mining machine that excavates the regolith to a depth of 3 meters. It separates particles 50 micrometers in size from the mass of regolith, heats them to 600 to 700 degrees centigrade, and collects the gases from this process in high-pressure cylinders. The processed dirt is then ejected back on the surface. Dirt collection is done by a bucket wheel excavator pivoted on an arm attached to the mining machine. The excavator sweeps an arc of 120 degrees as the machine moves slowly forward. The dirt is conveyed to coarse and fine sieves and the rejected portion is dumped over the side of the mining machine. The sieved soil is electrostatically beneficiated (separated) to amass particles 50 micrometers and less in size. This is the size measured to contain the highest helium content. These particles are conveyed to a heater, where they are heated to 700 degrees centigrade to boil off the trapped gases. When cooled, the particles are dumped into the trench dug by the excavator.

The source of the heating is sunshine, which is beamed to the Mobile Miner by a solar disk 110 meters in diameter mounted nearby. The solar disk tracks the Mobile Miner as it moves and beams the energy to a 10 meter receiver on the Miner. This energy is concentrated in an oven, where the gases are boiled off the sized particles of regolith. The gases are collected by a compressor, stored in cylinders, and transported to a central condensing station by automated ground service vehicles. The gases—including oxygen, hydrogen, nitrogen, carbon dioxide, and methane as well as helium—are then cooled from 300 to 55 degrees Kelvin in radiators. As each of the gases condenses, it drains into a vessel, where it is stored as a liquid. Helium at 55 degrees Kelvin is cooled and condensed further in a device called a cryogenerator to 1.5 degrees Kelvin to separate the isotope, helium 3, from helium 4. Separation is contrived by means of a superleak system that allows liquid helium 3 to drain off separately. During the condensation process, other volatile elements and compounds are stored, including water, methane, oxygen, nitrogen, and hydrogen. These volatiles are useful for life support for the lunar base and mining operation.

The Mobile Miner is designed to collect 33 kilograms of helium 3 a year. The machine excavates 1258 tons of soil an hour, of which 566 tons are processed. The machine's power consumption is figured at 30 kilowatts, based on scaling, from the Cleveland Trencher Model 8700, which is capable of excavating 1900 tons of soil an hour on Earth. The Mobile Miner for the Moon would have a mass of 18 tons and use five conveyor belts. It would excavate a trench 3 meters deep and 11 meters wide, moving along at 23 meters an hour. Operating power required would be 200 kilowatts. Energy required for processing the 556 tons per hour would be 12.3 megawatts. In one year, working 3942 hours, the machine would excavate a volume of lunar soil equal to one square kilometer. Because solar heat would be used to extract the volatiles from the soil, the Mobile Miner would be operated only in daylight, a span of about 14 days.

Mining the Moon could be a profitable enterprise. The scientists estimate the profit in terms of energy payback from helium 3 would run as high as 266 to 1 over 20 years. They concluded, "Preliminary investigations show that obtaining helium 3

Figure 7.2. Oceanus Procellarum on the Moon as photographed by Lunar Orbiter II Nov. 25, 1966. Outfall of helium 3 from the solar wind is believed to be concentrated in the lunar maria. (NASA).

from the Moon is technically feasible and economically viable."

A CAVEAT

As mentioned earlier, estimates of the energy payback of lunar helium 3 tend to run quite high. The Solar System Exploration Committee of the NASA Advisory Council reported that mining the Moon for helium 3 would yield an energy payback ratio of 250. This figure, also cited by the Fusion Institute, could be compared to 20 for uranium light water reactors and 16 for coal, the council reported.[12] These estimates lend the Moon an economic allure previously unsuspected. The prospect of helium 3 as fusion fuel is not new and the use of the deuter-

ium-helium 3 fuel cycle in space fusion reactors was speculated in a research study at NASA's Lewis Research Center, Cleveland, in 1962 and elsewhere. However, the University of Wisconsin group's 1986 paper was the first to cite lunar helium 3 as a fusion resource. What is new about this prospect is the realization that there may be an abundance of astrofuel on the Moon, as the data developed by the Fusion Technology Institute scientists at the University of Wisconsin suggested in 1986. Analyses of lunar samples showed that the soils of maria basins were enriched in the titanium ores, which, as mentioned earlier, have the highest helium content.

As charted by Institute scientists, the maria, comprising 20 percent of lunar surface, may contain 600,000 metric tons of helium 3, whereas the

Figure 7.3. Artist's visualization of the automated mobile lunar mining machine described by Igor Sviatoslavsky and Mark Jacobs. The machine is designed to excavate lunar soil to a depth of 3 meters, separate 50 micron particles from the rest, heat them to 700 degrees C, collect the evolved gases in pressure vessels and dump the spoil back on the surface. (University of Wisconsin)

highlands and basin ejecta, comprising 80 percent of the surface, may contain 500,000 metric tons. These estimates are based on assessments of helium content of the lunar samples.[13] Thus, the systematic sampling and exploration of the high-tita- nium soils should have a high priority on future lunar missions, according to Eugene N. Cameron, of the University of Wisconsin Center for Space Automation and Robotics. Cameron cited the prospects of finding helium 3 in the Mare Tranquillita-

Figure 7.4. With Conrad and Bean in Oceanus Procellarum. Mission Commander Charles Conrad, Jr. aligns the antenna on the central station of the Apollo 12 lunar experiments package Nov. 19, 1969. Soil samples from this and five other landings revealed the presence of the isotope, helium 3 in the lunar soil. The footprints in the foreground are those of Alan L. Bean who is taking the picture. Overhead, Richard F. Gordon, Jr. orbits the Moon in the Apollo Command and Service Module. Conrad and Bean lifted off the surface in the Lunar Module to rejoin Gordon in Apollo, bringing 75 pounds of lunar soil with them. (NASA)

POWER FROM THE MOON

tis, the region of the Apollo 11 landing. Tranquilli-tatis has an area of 190,000 square kilometers (73,360 square miles) and seems likely to contain regolith averaging 30 ppm of helium.

If only 30 percent of the area is mined to a depth of 10 ft and recovery from mining and processing is 60 percent, Cameron estimated that about 1700 metric tons of helium 3 could be collected.[14] Cameron carefully pointed out that the data on which the potential recovery of helium 3 is based are limited to the Apollo expedition samples (833.9 lb), the Soviet samples (hardly a kilogram), and remote sensing. Only a tiny fraction of the lunar maria surface was sampled, and no area has been sampled systematically, he said. Remote sensing maps based on gamma ray spectroscopy and reflectance measurements have insufficient resolution for mining site selection, he said.

As prelude to helium mining, Cameron's "caveat" warned, "deficiencies in present information" must be remedied by systematic exploration and sampling, by defining minable portions of the regolith, and by estimating tonnage and helium content. "Given the enormous potential of lunar helium as a source of energy, such work should have a high priority in future lunar missions," Cameron said.

Institute scientists noted that in 1986, the United States spent more than $40 billion for coal, oil, gas, and uranium to generate electricity. At $1 billion a ton, the 25 tons of helium 3 that could have met this demand would have been a bargain. On the basis of those 1986 costs, helium 3 would have a value of nearly $2 billion a ton. "At that rate, it is the only thing we know of on the Moon that is economically worth bringing back to Earth," the Institute group said.[15]

ASTROFUEL IN SPACE

The prospect of helium 3 as a high-energy fuel for space vehicle propulsion has been examined with enthusiasm by scientists of the Fusion Institute. Very high temperatures of the fusion reaction in magnetic confinement—up to a million times those of chemical propellants—greatly increase propel-lant efficiency. As mentioned earlier, the electrically charged particles released by the fusion reaction can be steered through a rocket nozzle by magnetic fields to provide thrust. The exhaust of hot plasmas can produce a specific impulse of tens of thousands of seconds—tens of thousands of times those of chemical propellants.* John F. Santarius, senior scientist and group leader in Plasma Engineering at the Institute, cited typical plasma burning temperatures of 500–1200 million degrees Kelvin. Exhausted through a nozzle, plasmas at such high temperatures can provide a range of specific impulses of 50 to 1 million seconds.[16] Some would be too high for missions confined to the Solar System but would be useful for missions beyond. Santarius and his colleagues said that specific impulses could be controlled to fit specific missions and could be reduced from 100,000 to 200 seconds by adding matter to lower the temperature of the exhaust plasma. Santarius predicted that when fully developed, helium 3 fusion propulsion will dominate future exploration and development in the Solar System. Fusion rockets might be available in the early part of the 21st century, he said.

In terms of a standard ship payload, helium 3 propulsion could reduce the flight time from Earth to Mars from 260 days by chemical propellant to 80 days and the flight time to Jupiter from 1000 days to 240 days. At the same time, it would allow double the cargo from Earth to Mars and six times the payload to Jupiter.

Helium fusion technology is seen as the answer to routine interplanetary flights serving settlements or stations on the Moon and Mars. It would provide the movement of supplies and personnel to scientific outposts, observatories on the Moon, and settlements on the Moon and Mars. Ultimately, it might provide the energy to divert resource-rich asteroids to orbits within reach of the Earth, Moon, and Mars.

The technical aspects of lunar helium 3 as an energy source in the 21st century have been explored in some depth and are fairly clear. But the

---

*Specific impulse is a measure of propellant efficiency in terms of seconds. It is expressed as a number of seconds obtained by dividing thrust in pounds times burning time by propellant weight. Specific impulse values of chemical rockets range from 220 seconds (alcohol) to 375 seconds (liquid hydrogen).

legal and political aspects of extracting this material from the Moon by national or international entities for their own use or profit remain vague.

There is another way of obtaining helium 3 that antedates the scenario that a billion metric tons of the isotope might be mined on the Moon. It was proposed by the late Krafft A. Ehricke (1917–1984), a leading engineer–theoretician of modern space travel. In 1984 he described a process for obtaining helium 3 from the decay of excess tritium produced by the deuterium–tritium reaction in nuclear fusion plants. He believed that these plants should be built on the Moon, where public concern about radioactive tritium would not inhibit their development. The helium 3, as a by-product, could then be shipped to Earth for radiation-free power generation in deuterium–helium 3 fusion reactors.

In 1984 Ehricke projected a rationale for developing the Moon for human settlement, commerce, and industry as an annex of the Earth at the First National Conference on Lunar Bases and Space Activities of the 21st Century. At that time the feasibility of mining the isotope on the Moon had not been widely considered.

Ehricke was one of the youngest of the German engineers who had worked on the development of the V-2 rocket for the German army at Peenemunde under Wernher von Braun. With other engineers on the von Braun team, he emigrated to the United States after World War II to work with the U.S. Army on rocket development. A 1942 graduate of the University of Berlin in aeronautical engineering, Ehricke had been assigned to the V-2 propulsion system for three years at Peenemunde. The U.S. Army employed him as a jet propulsion engineer at Fort Bliss, Texas, from 1947 to 1950 and as chief of the gas dynamics section, Army Ballistic Missile Center at the Redstone Arsenal. He later became program director of the Centaur rocket, the first to use hydrogen for fuel, for General Dynamics Corporation. In 1968 Ehricke became chief scientist for space systems and applications for the space division of the North American Rockwell Corporation and continued in that role until his retirement in 1976. He then organized his own consulting company, Space Global, Inc. One

of the highlights of his career was being named to the International Aerospace Hall of Fame in 1966.

A POLYGLOBAL CIVILIZATION

Ehricke's vision of development on the Moon was grand in scale and minute in detail. Essentially, he viewed pioneering efforts to establish a base on the Moon as an evolutionary step toward the birth of what he called a "Polyglobal" civilization. It would free humankind of the limits to growth imposed by confinement to a single planet. To sum up his views on the dynamics that seemed to him to be the driving force of space exploration and expansion, Ehricke coined the phrase, the "Extraterrestrial Imperative." It implied that human civilization was fated to expand throughout the Solar System and beyond. The key that unlocked the gate was the Moon.

In many respects, Ehricke's ideas are as revolutionary today as those of Konstantin Tsiolkovsky were nearly a century ago, although Ehricke stood on the shoulders of Tsiolkovsky, who stood on the shoulders of Galileo and Newton. Tsiolkovsky invented space travel and Ehricke projected it as the means of establishing an interplanetary civilization.

Ehricke, who died in 1984, buttressed his dreams with engineering designs and economic rationales. He projected not only what colonists could do on the Moon, but how they could do it—how they could make a living there. In his view the Moon was the ladder to the cosmos, the anvil on which an interplanetary, and eventually interstellar space transportation system would be forged.

In Ehricke's view the logical power source for lunar occupation would be deuterium–tritium fusion power plants, yet to be developed in the power economy of Earth. Deuterium and tritium are heavy isotopes of hydrogen. When they are fused by heating to millions of degrees (in magnetic containment), the reaction produces helium, neutrons, and energy. The reaction is clean compared to fission but must be shielded because of radioactive tritium.

POWER FROM THE MOON

The Moon is an ideal site for the development of a deuterium–tritium (D–T) reactor, Ehricke maintained, because the lunar vacuum facilitates construction of the reactor plasma chamber without restricting its size. The chamber may be as large as a football field. The lunar environment simplifies reactor maintenance, Ehricke said, and facilitates the use of superconducting magnets for magnetic containment.

Excess tritium bred with the reaction can be stored on the Moon, decaying over a period of 12 years into the extremely valuable isotope helium 3 (plus an electron), Ehricke said. As related earlier, the value of helium 3 lies in its use as a fuel for a "super power plant—a deuterium–helium 3 fusion reactor." But because of the high temperature required to ignite the D–He plasma, deuterium–helium 3 fusion is difficult to attain. Still, it is important for fusion technology on Earth because the reaction is clean: no radioactive tritium is involved. He and other researchers said that only 7 percent of the released energy is carried away in neutrons, so that virtually no radioactive isotopes are generated.

Ehricke's projection of lunar power in 1984 did not consider the feasibility of mining solar wind deposits in the lunar regolith, a scenario presented at the second conference on Lunar Bases and Space Activities of the Twenty-First Century four years later. Ehricke believed that helium 3 could be produced in deuterium–tritium fusion plants on the Moon. The fusion plants could not only provide power for lunar industry and colonies but export helium 3 to Earth.

A trading partnership based on nuclear fusion technology could grow up between the Earth and the Moon. Earth would provide deuterium and lithium (a source of tritium) and the Moon would supply the helium 3 as a by-product of the deuterium–tritium fusion power plants. Ehricke calculated that 84 tons of helium 3 would generate 500 gigawatts (billion) a year in deuterium–helium 3 power plants on Earth at a profit of $44 billion a year from a net of one cent on each kilowatt-hour. The 84 tons of helium 3 could be produced in 12.3 years, the half-life of tritium, from the decay of 168.9 metric tons

of excess tritium produced in the lunar deuterium–tritium fusion plants.

Ehricke had been considering helium 3 for several years. In September 1982 he reviewed its application to space ship propulsion in a paper delivered at the International Symposium on Space Economics and Benefits, sponsored by the 23d Congress of the International Astronautical Federation in Paris. There he said that the deuterium–helium 3 reaction is well suited to a steady-state fusion drive for interplanetary and possibly cislunar space vehicles.

However, he warned of a problem in cislunar space. The proton and alpha particle exhaust of the fusion drive could become trapped in the Earth's magnetic field as the spacecraft passed into the radiation belts. The trapping would gradually increase the charged particles in the belt and make it more hazardous for humans to cross. This effect also would create a harsher radiation environment for satellites in the belt.

Ehricke was convinced that there would be a big demand for helium 3 in the future, for power and propulsion, and that the Moon was the place to develop it. He calculated that 62 percent of the construction material for a magnetic confinement nuclear power plant could be provided by lunar resources. "The deuterium–tritium fusion reactor can produce twice as much tritium as you need," he said.

The overage decomposes in 12.3 years into helium 3, which is immensely valuable, a terrific propulsion system. Also, helium 3 with deuterium goes into helium 4 and protons. Those protons I combine with electrons and I have hydrogen. By building up helium 3 I am building up water.

Then after 20 years I'm in possession of a thousand tons of helium 3 and almost 400 tons of water. Now I'm truly independent. That technology I can use anywhere in the Solar System.[17]

INDUSTRIALIZATION

The prospect of mining the Moon for oxygen, hydrogen, and helium 3 offers an initial basis for the

1 4 9

industrial development of the Moon. It put a fine edge on the proposition that the Moon is a natural annex to Earth's industrial civilization as well as the gateway to the exploitation of Solar System resources. The availability of hydrogen from the regolith sharpened this perception in the 1980s. Earlier, scenarios of lunar development had assumed importation of hydrogen from Earth, unless ice was found in the permanently shadowed recesses of the polar regions. In theory, at least, the hydrogen problem was resolved by the belief that more than enough could be extracted from the soil to supply the needs of lunar habitation and industry.

With oxygen and hydrogen, a lunar base was in a fair way to become self-supporting. At least the prospect appeared reasonable. The helium 3 scenario added the El Dorado motif to the more prosaic economic incentives for going back to the Moon. Krafft Ehricke told the first Lunar Bases conference that the industrialization of the Moon had to pay off in classical economic terms to be viable. He insisted that the Moon had to be developed by private enterprise. Otherwise, he said, the rise of lu-

nar civilization as a trading partner of Earth would never be realized.

These ideas were shared widely in the techno-scientific community based in NASA, the aerospace industry, and universities and government laboratories. Ehricke had cited three technologies that would attract private investment. First was the development of nuclear power plants. Second, was the extraction of lunar oxygen for lunar and cislunar space applications. Third was the use of lunar materials for construction, shielding, commerce, industry, and agriculture.

From resources mined on the Moon, lunar industry could produce sheet metal and trusses of aluminum, magnesium, titanium, iron, or alloyed metals. The production list was long: castings, bars, wire, powders, glass and glass wool, ceramics, refractories, fibrous and powdered insulation, ceramics, insulation, conductors, anodized metal, coatings, thin films, silicon chips, photovoltaic (solar) cells, and thermal and radiation shielding for use in spacecraft, space stations, and free-flying laboratory modules. A lunar industry could produce commercial communications and solar power sat-

Figure 7.5. Artist's visualization of a science outpost on the Moon. The base consists of an assembly of space station size modules. The power source would consist of photovoltaic panels to transform sunlight into electricity and an array of nuclear power units. (Pathfinder).

ellites. Once the lunar industrial infrastructure was in place, satellites produced on the Moon could be put in orbit from there more cheaply than from Earth. Over the long term, the saving in propellant costs would be significant, in the opinion of some experts, significant enough to justify satellite manufacturing on the Moon.

Of fundamental importance to lunar industrial development is a mining technology that works in the lunar environment. There are special considerations on the Moon, according to a study by E. R. Podnieks and W. W. Roepke, U.S. Department of the Interior, Bureau of Mines.[18] They cited tests showing that surface friction and rock strength are higher in hard vacuum on the Moon than on Earth. This may be due to greater adhesive forces (in vacuum) and lack of moisture. Adhesion forces on tools and rock surfaces will create a problem with chip formation and clogging during drilling. For example, dust adhesion to space suits and tools can be seen in *Apollo 16* photos, the authors said.

During the *Apollo 15* mission to the lunar Apennines–Hadley Rille region on the Moon in the summer of 1971, David R. Scott and James B. Irwin experienced drilling problems. They tried to insert a heat probe to a depth of 10 feet into the regolith at the foot of Apennine Mountains to measure heat flow from the interior. They used a jackhammer drill to make the hole but found that the soil was so resistant that the drill would not go in far enough. It penetrated to only 4.2 feet at one site and 3.3 feet at another before the drill stem broke. The Bureau of Mines investigators reported that friction in the lunar regolith may range up to 60 times frictional forces in terrestrial soil.

Podnieks and Roepke explained that since the lunar surface consists of degassed, pristine materials, excavation and mining can be expected to encounter high friction unless some means, such as a gaseous lubricant, is found to minimize it. Drilling equipment designed to reduce friction was provided for the *Apollo 16* mission to the Cayley–Descartes formation. One of the chores assigned to John W. Young and Charles M. Duke, Jr., was to emplace a pair of heat probes in holes 10 foot deep. Duke drilled one hole successfully and inserted the fiberglass probe, but before the second hole was

drilled, Young caught his boot in the cable and yanked it off the installed probe, precisely at the connector. Duke used the drill again to extract a deep core sample.

A number of investigators have suggested that blasting with high explosives may be an efficient method of releasing gases and separating ores on the Moon. The Bureau of Mines conducted tests on the performance of explosives in a hard vacuum. The behavior of blast fragments was similar to that on Earth, it was found. Explosives are available that will work in the lunar environment, the bureau reported.

Nuclear explosives have been considered by Ehricke and others. Because of the lack of water, nuclear explosive excavation would not produce steam, as it does on Earth. Without steam to crack cavern walls, the nuclear detonation would produce impermeable glass-lined walls as the melt solidified. Alternatively, an electrothermal technique of fragmenting rocks would be effective, the Bureau of Mines reported. Lasers might be used for breaking rock. A multidisciplinary approach to lunar mining technology was needed, the Bureau advised.

THE AMES STUDY

Much of the preliminary laboratory and theoretical work to define the feasibility of lunar mining was done during the Apollo years and immediately thereafter, when the occupation of the Moon seemed a logical next step. These surveys were filed away when NASA retreated from the Moon and concentrated on low Earth orbit with the shuttle. Some of the surveys have reappeared to support the renaissance of interest in the Moon.

A detailed study by scientists at NASA's Ames Research Center in 1973 concluded that it would not be economically feasible to mine, refine, and ship lunar materials to markets on Earth. The study noted that a transportation system based on chemical or nuclear powered vehicles would be too costly.[19] The use of lunar materials could be rationalized only on the Moon or in orbit.

The study contained a pioneering proposal for

development of an electromagnetic accelerator to boost material into lunar orbit instead of lifting it by means of nuclear or chemical rockets. The system, as described earlier, was composed of dual coils—a stationary coil, which included the track and a moving coil which would accelerate down the inside of the track until a velocity of about 2900 meters per second (6487.30 miles per hour) was reached. At that point the payload would separate from its carrier and enter orbit. The report calculated that a track 40 kilometers (24.8 miles) long would be necessary to achieve orbital velocity under 10-g constant acceleration. The payload was estimated at 7 metric tons and the cost of building the accelerator at $40 billion. Despite these formidable estimates, the idea of hurling material off the Moon by electromagnetic acceleration has become conventionalized in recent lunar development scenarios.

POWER

All scenarios of lunar development require a power source, solar or nuclear. Because of the two-week-long lunar night at all except polar latitudes, nuclear power is preferred in most projections. As a result of research and development efforts in the 1960s, it is available for lunar or space stations. However, its transportation off the Earth presents safety concerns.

Two types of nuclear power systems were developed for space or lunar application by the Atomic Energy Commission and its successors (the Energy Research & Development Administration and the Department of Energy). The system that has been used successfully in providing electric power for satellites and experiments on the Moon and Mars is the radioisotope thermal electric generator (RTG).

In this system heat released by the decay of plutonium 238 is converted to electrical energy by a bank of thermoelectric elements. When heated, these elements emit electrons. This type of generator was first flown aboard a Navy Transit navigation satellite in 1961. It was developed as a System for Nuclear Auxiliary Power (SNAP) and various

Figure 7.6. In addition to providing electric power, the photovoltaic array on the lunar surface shown here would be useful in testing various types of ceramic materials for power conversion. Its primary utility would be on the Moon but application to Mars also is being studied. (Pathfinder).

models have powered weather, communications, and military communications satellites as well as scientific experiments emplaced on the Moon and Mars. A SNAP 27 RTG provided the power for scientific stations set up on the Moon by the crews of Apollo 12, 14, 15, 16, and 17. Two SNAP 19 generators provided power for the experiments on the Viking landers on Mars. Lightweight and long-lived, the SNAP 19s produced 35 watts each at 4 volts for more than two years. They were only 15 inches long, 23 inches across the housing at the base, and each weighed 34 pounds.

The more powerful source of nuclear energy is the uranium-fueled reactor, designed for rocket propulsion. It heats a gaseous propellant such as hydrogen to produce thrust and can also be applied to run a turbine generator to produce electricity. The first administrator of the old Atomic Energy Commission, Glen T. Seaborg, described a flyable reactor not much larger than an office desk and capable of producing 1500 megawatts of electrical power (the power of the Hoover Dam, he said).

High-speed pumps move three tons of hydrogen (liquid at −450 F) past the reactor's white hot fuel elements at 4000 degrees F. Seaborg predicted that the nuclear rocket engine would have twice the specific impulse (pounds of thrust per pound of propellant flow per second) of the most efficient chemical engines. It would make possible long-distance voyages in space, he said. Such an engine had been visualized as early as 1955 by a program called Rover.

Expectations for nuclear-powered spacecraft were high before the rapid expansion of domestic nuclear power plants and the radiation hazards associated with them aroused public concern about the safety of this technology.[20] In 1969 the Atomic Energy Commission completed ground tests of a system called NERVA (Nuclear Engine for Rocket Vehicle Application). NERVA operated at 1500 megawatts (thermal) and produced 75,000 pounds of thrust. It was designed as an upper-stage engine, to be boosted off the ground by chemical rockets and ignited above the atmosphere. Even though this process was expected to avoid contamination of the biosphere by radioactive exhaust, the possibility of a first-stage failure allowing the nu-

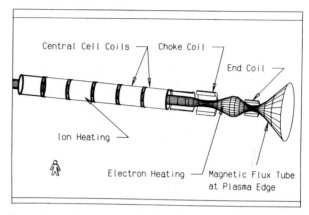

Figure 7.7. One type of nuclear fusion rocket. In this magnetic configuration, plasma is contained within magnetic flux tubes and exhausted to produce thrust. (College of Engineering, University of Wisconsin).

clear stage to crash at the launch site or in the ocean led to a ban on flight testing it, and NERVA never flew. Although NASA's Space Nuclear Propulsion Office had proposed that the Los Alamos Laboratory should design and build a 5000 megawatt reactor as a prototype for a more powerful engine, the nuclear space engine development project was halted late in 1969.

There was concern also about launching SNAP generators containing several pounds of plutonium, even though they were heavily shielded. When *Apollo 13* was forced to abort its lunar landfall in 1970 and take up a free return trajectory to Earth, the SNAP 27 generator fell into the Pacific Ocean with the Lunar Module. The LM was not jettisoned at the Moon but instead remained attached to the Apollo Command and Service Modules to provide oxygen for the crew after the bulk of their supply of oxygen was lost with the explosion of the service module oxygen tank during the outbound journey. Just before reentry, the crew detached the Command Module from the Service and Lunar Modules which plunged into ocean, SNAP 27 and all.

The use of nuclear reactors on the Moon presents a transportation safety problem that has not yet been resolved. Such a device, like the NERVA engine, would probably be based on a reactor called KIWI, which was developed about 1959 by the Los Alamos Scientific Laboratory and tested at its Nevada Test Site, Jackass Flats. The reactor developed 100 megawatts and an advanced version of it, KIWI-B, reached a power level of 1000 megawatts (thermal) energy.

A nuclear power technology that could provide all the power requirements of a lunar base has been available for 20 years. The problem is getting it from the Earth to Moon in a manner that allays public safety concerns.

WHAT THE MOON IS MADE OF

The chemical composition of the lunar regolith is known from samples at nine sites, six Apollo and three Soviet Luna, and from remote sensors flown in lunar orbit by *Apollo 15* and *16*. As defined by

James R. Arnold (professor of chemistry, University of California, San Diego), the principal classes of silicates are represented by the dark-colored mare basalt, the light-colored highland material, and a rock complex called KREEP, an amalgamated mixture of potassium, rare earth elements, and phosphorus. The highlands have a higher percentage by weight of aluminum and calcium than basalt, which is richer in iron and titanium.

The ground tracks of *Apollo 15* and *16* were chemically scanned as these vehicles circled the Moon while two crewmen were on the surface. The scanners were a gamma ray spectrometer and an x-ray detector. The spectrometer was built around a scintillation detector that recorded photons in the range of 0.3 to 10 million electron volts. The detector responded to radioactive thorium, potassium, and uranium, as do similar instruments on Earth. Another sequence of responses is produced by the interactions of cosmic ray–generated neutrons with major elements in the regolith. This allows measurements of oxygen, silicon, iron, and magnesium.

The x-ray detector consisted of proportional counters with filters that recorded secondary x-rays emitted from the surface, excited by the solar x-ray continuum. The records appeared as ratios among the element abundances of magnesium, aluminum, and silicon. Ground truth—that is, verification of these results from the scanners—was provided by sample analysis from the landing sites. A region near a feature called Van de Graaff in the southern highlands of the far side of the Moon showed a significant concentration of thorium and iron. Was that an ore body? The gamma ray detector had a resolution of 50 kilometers and the x-ray detector, 10 kilometers. Additional chemical information was provided by the reflectance spectra of the soil at visible and near infrared wavelengths, Arnold said. By this means the mineral pyroxene containing iron and magnesium could be readily identified.

In the maria the presence of dark glass and opaque minerals high in iron and titanium had a strong effect on the reflectance spectrum, and weaker spectral features indicated the mineral olivine and free metallic iron. These observations of reflectance spectra can be made from Earth-based telescopes with a resolution comparable to that of photographs. There have been hints of ore bodies in addition to the apparent abundances of iron and thorium on the far side. The green glass found at the foot of the lunar Apennines by Scott and Irwin on *Apollo 15* and the orange glass found by Harrison H. Schmitt and Eugene Cernan at the foot of the Taurus Mountains on *Apollo 17* suggested the presence of titanium ores, according to Arnold. Larger areas of similar material have been located from orbit by distinctive color, Arnold added. One

rock (No. 12013) brought back for analysis had a much higher concentration of radioactive elements than the average specimen. "Still," said Arnold, "we have seen nothing like the native copper of Michigan or the tin mines of Bolivia."

TREATIES

Whatever the mineral wealth of the Moon may be, existing international treaties require (in principle) that it be shared among all the peoples on Earth.

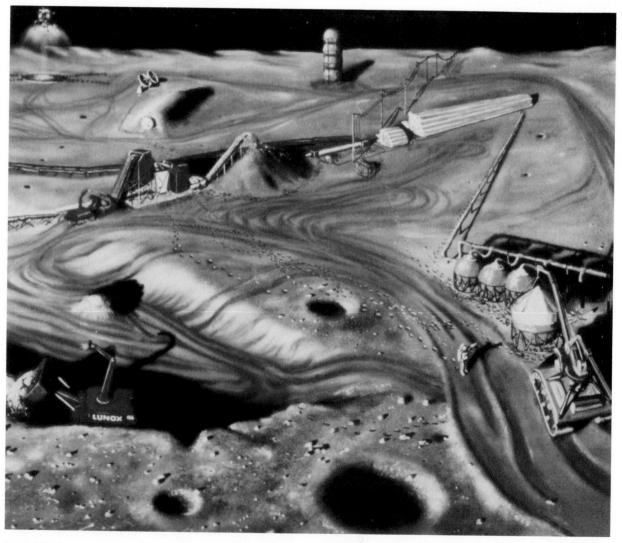

Figure 7.8. A processing plant on the Moon would provide the means of extracting oxygen, hydrogen, helium and other elements from the rocks and soil. Oxygen would be obtained directly from the rocks. Hydrogen and helium, deposited from the solar wind, would be collected by heating the lunar fins. This artist's rendition of a lunar oxygen mining camp illustrates one phase of anticipated lunar development. Project Pathfinder (NASA)

How this could be done in the context of national lunar settlement programs is not resolved.

A prohibition against national appropriation of territory or resources on the Moon was set forth in the 1967 Treaty on Principles Governing the Activities in the Exploration and Use of Outer Space, including the Moon and other Celestial bodies. This document, sometimes referred to as the Outer Space Treaty or the Principles Treaty, has been ratified and its principles accepted by the space-faring powers (United States, the U.S.S.R., European Space Agency nations, Japan and China) as well as by other states.

The prohibition against national appropriation is specifically expressed by Article 11 in another treaty, the Agreement Governing the Activities of States on the Moon and other Celestial Bodies (the Moon Agreement), which entered into force July 11, 1984. By the end of 1985 it had been ratified by Austria, China, the Netherlands, the Philippines, and Uruguay, but not by either the United states or the Soviet Union.

The Moon Treaty allows use of lunar material for scientific purposes (such as the rocks returned by Apollo) and for technological development, but claims to a portion of the Moon or its resources are outlawed by both treaties. According to an aerospace law consultant, Amanda Lee Moore, the principle of nonappropriation as stated in the Principles treaty is "sufficiently normative in character so as to be considered a valid principle of international law both in treaty and custom."[21]

In a symposium paper on Extraterrestrial Law and Lunar Bases, Harrison H. Schmitt, the former astronaut and U.S. senator from New Mexico, and Christopher C. Joyner, Department of Political Science, George Washington University, noted that there is no rule precluding profit from or private enterprise in space. International space law does not prohibit commercial activities in space and recognizes that they will take place, with the state responsible for them.[22] They noted also that the present rationale for a lunar base refers to its use as a scientific and as a logistics base for planetary exploration. These are activities accomplished with government support.

Conceivably, a lunar facility could be established and lunar resources extracted for their component materials on a profit-making basis. Other activities could be started using the Moon as a base without further legal notice to an international body. Space law, they said, requires little beyond peaceful use in accordance with international law on a nondiscriminatory basis, with government authorization and supervision and the dissemination of results.

Schmitt and Joyner suggested an international regime like Intelsat (the International Telecommunications Satellite Organization) for the governing of international activities on the Moon. The agency could be called "Interlune," they added. It would be comprised of an assembly of parties, a board of governors, a board of users and investors, and a director general. It would provide cooperative international management of a lunar base for the benefit of member states, users, and investors. In all, five agreements on the behavior of nations in space and on the Moon and planets have been generated in the United Nations.

In addition to the Principles and Moon Treaties, a treaty providing for the rescue and return of astronauts and the return of objects launched into space was put into effect in 1968. It was followed in 1972 by a convention on international liability for damage caused by space objects and in 1974 by a convention on the registration of objects launched into space.

The Moon Treaty was submitted to the United Nations Committee on the Peaceful Uses of Outer Space in July 1979 while negotiations were in progress on the inner space front dealing with the law of the sea. The Moon treaty raised issues that were being discussed in the law of the sea meetings, according to the report of the U.S. Senate Committee on Commerce, Science, and Transportation.[23]

Concern that the Moon treaty would interfere with the exploitation of extraterrestrial material by U.S. private industry was expressed by two senators, Howard W. Cannon (D–Nev.), the committee chairman, and Adlai E. Stevenson (D–Ill.), chairman of the Subcommittee on Science, Technology, and Space. They wrote to Morris K. Udall, chairman of federal Technology Assessment Board, that "it is being argued that the agreement if ratified could inhibit or even prohibit" industry from developing or exploiting extraterrestrial material. They asked that the Office of Technology Assessment examine the impact of the treaty on the use of such material.

The Senate committee report stated that NASA believed that the use of lunar resources was not justified unless a program of the dimensions of a solar power satellite system depended on it. NASA studies, said the report, emphasized that lunar resources provided opportunities for conducting large-scale operations in space. The report noted that the role of private enterprise in the development, ownership, and operation of space systems (principally communications) exceeded $3 billion in the period 1960–1980. New entities were seeking to enter the commercial arena in space. Already, the report said, technical bodies like the International Telecommunications Union had begun to feel the influence of the "North–South debate" involving the interests of developed countries in conflict with those of developing countries.

The committee cited seabed mining negotiations on the rights of private enterprise to mine the seabed. During six and one half years of negotiations under the aegis of the third United Nations Conference on Law of the Sea (UNCLOS III), commercial interests had invested several hundred million dollars in locating, mining, and processing seabed resources. But full-scale deep-sea mining by American companies could not occur because draft treaty texts required modifications before they could serve as a basis for private investment acceptable to these companies, the committee reported. "Apparently," said the committee report, "no major U.S. seabed mining company presently contemplates making the large scale investment in actual recovery facilities so long as the extent to which that investment will be respected under future law of the sea regimes is in doubt. . . . Even after 10 years of delay because of the politics involved, the U.S. government appears unprepared to exercise the unilateral right of seabed exploitation it consistently has claimed to have."

In several respects, the committee considered the Moon treaty as an extraterrestrial analogue of the seabed mining treaties and their difficulties.

The committee observed that the Moon Treaty substantially reiterated the Principles Treaty but went somewhat beyond it in extending constraints on the freedom of states to use the Moon and other celestial bodies.

The Moon Treaty had a curious, uncertain history. It was proposed in 1971 by the Soviet Union, which submitted a draft to the United Nations dealing only with activities on the Moon and space around it. A U.S. draft in 1972 presented the idea of advance international notification of missions to the Moon and other celestial bodies. It proposed also that the Moon and its natural resources be the "common heritage" of mankind.

The "common heritage" bequest was rejected by the Soviet Union as "premature politically and juridically vague and lacking in specificity." Then, in 1973, the United States rejected a proposal by India that exploitation of lunar resources await adoption of a treaty that would govern it. The United States also announced it opposed a moratorium on exploiting lunar resources. As it became apparent that lunar resources represented a vast reservoir of real wealth, the notion that they were the "common heritage of mankind" became moot. Although it had entered into force under U.N. rules, the Moon Treaty was not ratified by the only two powers it would affect.

Two Legal Concepts. At the second conference on Lunar Bases and Space Activities of the 21st Century, in 1988, Joanne Irene Gabrynowicz of the Space Studies Department, University of North Dakota, enlarged the discussion of the U.N. treaties on the exploitation of the Moon.[24] She pointed out that the "common heritage of mankind" and the "province of all mankind," while fundamental legal concepts in international space law, may be interpreted differently in the application of each concept to lunar development. This would appear to be an important consideration in view of accumulating evidence of energy and material resources on the Moon.

Since the formulation of these concepts (the so-called "mankind provisions"), controversy has arisen about their meaning as applied to a nation's right to explore and use a common environment, such as space or the seas, and a nation's obligation to share its bounty.

The "province of all mankind" concept in Article 1 of the 1967 Outer Space Treaty refers to activities such as exploration and use on the Moon and other celestial bodies while the "common heritage" provision of the Moon Treaty of 1979 refers to material objects. This distinction, cited by a Soviet expert, Boris Maiorski, would allow the United States or the U.S.S.R. to follow its own policy on distributing wealth from space materials, Gabrynowicz said. The author pointed out that the "common heritage" provision does not appear in the 1979 treaty, which neither the United States nor the Soviet Union have ratified. Therefore, said the author, applying the distinction between "activities" and "materials" to the 1967 treaty, which both nations have ratified, would allow the United States to support the treaty "without relinquishing its conviction that private enterprise in space ought to be profitable. . . ."

THE ANTARCTICA CASE

As discussed earlier, the exploration of the Moon has been considered analogous to the exploration of Antarctica, legally as well as physically. Both regions are remote, are difficult to reach, and have hostile environments. Both have scientific importance and economic potential. Their early exploration was the product of international rivalry, and both have generated concepts of international law that apply to the expansion of human activities and settlements.

There is an important difference, however, at least now. Antarctica is the locus of conflicting territorial claims which must soon be resolved. Otherwise, conflicts are likely with the development of resources. The Falklands war may be a warning.

The first antarctic lands were sighted in 1820 by British, American, and Russian seamen, but the continentality of the land that lies beneath the ice cap was not established until early in 1840. Lieutenant Charles Wilkes, of the U.S. Navy, sent to the antarctic to establish an American presence in

the region, found a rocky coast in the Australian quadrant of the ice cap on January 19, 1840.[25] He assumed that it was continental land and called it the Antarctic Continent.* The extent of the land that lay beneath the ice was not known for another century.

The pace of antarctic exploration, characterized by teams of dogs pulling sleds and human muscle, accelerated with the introduction of the ice breaker ship, the tractor and the airplane into the antarctic environment. In preparation for the International Geophysical Year (1957–1958), the United States established seven stations on the ice cap, including Amundsen–Scott Station at the South Pole. The port of entry to Antarctica in that period was the U.S. Naval Air Facility on Ross Island, a volcanic spit on the coast of McMurdo Sound, an arm of the Ross Sea.

The Soviet Union established an inland station at a place called the Point of Inaccessbility. It was the region farthest from the coast of the continental ice sheet. National research stations were set up by New Zealand, Australia, France, Argentina, and Japan. By 1959, seven nations had staked out territorial claims in the antarctic. Sixty years of exploration by national expeditions had let to territorial claims by Australia, France, New Zealand, Norway, and the United Kingdom early in the century and later by Argentina and Chile.

The claims by Argentina and Chile, which conflict in some respects with Britain's claim and with each other, are based in part on the Papal Line of Demarcation, which divided discoveries in the Western Hemisphere between Spain and Portugal in 1494. The Line of Demarcation was defined by the Treaty of Tordesillas as the meridian 370 leagues (about 1175 miles) west of the Cape Verde Islands. All discoveries west of it were awarded to Spain; east of it to Portugal.

The claims by the seven countries cover four fifths of the ice cap, leaving one fifth, the region called Marie Byrd Land, with no claimant. The United States maintained its Little America base there during the International Geophysical Year and for several years thereafter. The base was es-

_____
* Wilkes Land.

tablished in 1928 by Rear Admiral Richard E. Byrd near the site of Roald Amundsen's old camp at the Bay of Whales. Members of Congress repeatedly sought to have administrations make a formal claim there, but none did. The American no-claim policy saved embarrassment, for years later soundings of the ice sheet showed that the land beneath Little America was sea bottom.

Following successful international scientific cooperation during the International Geophysical Year, the Antarctic Treaty was negotiated and signed by Argentina, Australia, Belgium, Chile, France, Japan, New Zealand, Norway, the Union of South Africa, the Soviet Union, the United Kingdom, and the United States. It was hailed internationally as an enlightened step toward the resolution of territorial disputes. Unique in the history of nations, the treaty set aside for benefit of science a continent as potentially rich as any other on Earth, even though its potential wealth was locked away under an ice cap two miles thick.

Economically, it appeared impractical to attempt to recover these resources as long as they existed in more benign environments. There was anthracite coal in the Trans-Antarctic Mountains. It had been found in 1907 by Sir Ernest Shackleton. Deposits of oil and gas could be assumed from plant and animal fossils from a time past when the continent now buried beneath the ice lay warm and open to the sun. The economic feasibility of exploiting antarctic resources may improve if world shortages of fossil fuels and critical metals become acute. In that event, national claims may be economically relevant.

The Antarctic Treaty, establishing a continent for science, free of nuclear testing, free of military bases and frontiers, open to any human being who wants to go there, has been considered a model for space treaties. But that may change. The treaty could be reopened for the adjudication of claims starting in 1989, when its 30-year moratorium on claims and disputes expires. At that time, any of the 12 signatories may call a conference of review and amendment.

Scientifically, Antarctica provided a key to the structure of the Earth, as the Moon has. Finds of coal, fossil trees, and the discovery of fossil bones of a dog-sized reptile in the Trans-Antarctic Mountains offered indisputable evidence of continental drift years before the concept of plate tectonics revolutionized geology. The antarctic cap is the analogue of the southern ice cap of Mars. But there the analogy seems to end. The Moon is the frontier of the 21st century; Antarctica of the 19th.

EIGHT

A NATURAL HISTORY OF MARS

As projected by the Presidential Space Task Group report of 1969, the next stop beyond the Moon was the planet Mars. Despite official dithering, delays, doubts, and cutbacks, that goal has never changed. The red planet has a long history in human cultures, antedating the space age by thousands of years. It was Nergal, a vengeful god, to the Chaldeans; Ares to the Greeks; Mars, god of war, to the Romans, a personification encoded in Western languages. Its color was no illusion. Its soil and skies are red.

From interpretations of telescopic observations in the nineteenth century arose scenarios of Mars as an abode of life.[1] This notion was enhanced by the appearance of lineaments on the surface of the planet that suggested the existence of channels.* The lineaments were seen in 1785 by a German astronomer, Johannes Schroter; in 1858, by Fra Pierre Angelo Secchi, papal astronomer, and particularly by Giovanni Schiaparelli of Milan in 1877. Schiaparelli called them "canali" and regarded them as waterways.

There arose from this perception the fantasy of a civilization struggling to survive the desertification of a planet in the process of desiccation by channeling meltwater from the polar ice caps to the temperate zones. A wave of darkening that appeared across the temperate zones in the Martian spring suggested growth of vegetation along the supposed canals.

The Mars of fantasy reached a peak in the late 19th and early 20th centuries in the science fiction of H. G. Wells and romances of Edgar Rice Burroughs that depicted an inhabited world. But the climax of this fantasy came in 1938 with the radio dramatizations of H. G. Wells' "War of the Worlds." The Martian invasion was dramatized by Orson Welles and the Mercury Theater as a sequence of radio news bulletins from the landing site in New Jersey. The effect was so realistic that police stations, radio stations, and newspaper offices were swamped with frantic queries from frightened listeners, despite assurances that the program was purely fictional.*

Mars loomed as a goal of the space age more than a decade before President Kennedy called for a manned lunar landing. In 1948 Wernher von Braun proposed a manned expedition to Mars by a fleet of space ships. Initially published in a West German magazine, the proposal, entitled "The Mars Project," was published as a book in 1952 by the University of Illinois Press and later issued as a paperback.†

Von Braun's study was a detailed engineering document. It described a flotilla of 10 interplanetary space ships with crews totaling 70 men. Each ship would be assembled in Earth orbit, to which three-stage ferry rockets would deliver the components, the personnel, and the propellant.

Once assembled, fueled, and ready to go, the vessels and their crews would depart Earth orbit and accelerate by means of chemical rockets on an elliptical path around the sun. At the maximum solar distance of the ellipse, the flotilla would approach the orbit of Mars at a tangent.

Essentially, this flight plan projected an orbital transfer strategy that remains standard. The space ships would fire their rockets for deceleration so that they would fall into an orbit around Mars. Three would be equipped with landing boats for the descent to the surface as winged gliders. The surface crew would return to Mars orbit in two of the boats, abandoning one on the surface. The entire crew would then return to Earth orbit in seven ships, leaving the three that had carried the landing boats—and the two boats—in Mars orbit.

Up to this juncture in the literature of space travel, interplanetary flight had been projected in terms of direct ascent from one body to another. "The Mars Project" appears to be the first detailed account that incorporated the concepts of interorbital transfer and orbital-rendezvous transfer.

Here for the first time the interplanetary vessel (or fleet) is assembled in and launched from orbit, an enormous saving in energy. It never lands but remains in space. When it reaches and establishes itself in the orbit of its target planet, the exploration party departs in landing boats, or "landers." For the return the party is propulsively delivered to orbit to make rendezvous with the circling mother ship. It returns them to Earth orbit, from which the crew is shuttled to the surface. The basic outlines of this strategy have not changed in 40 years. It is the way we went to the Moon (lunar orbit rendezvous) and characterizes all the current projections for manned Mars missions. The details, however, have changed considerably, as will be related later.

ILLUSION DISPELLED

Although engineering, navigation, and propulsion problems of a Mars mission appeared susceptible to solutions on paper (von Braun said he used a slide rule for his computations), the planetary environment was barely known, its nature obscured by fantasy and speculation. There was no evidence that a spacecraft could even land on the surface. The chemistry of the atmosphere was unknown, and its density was estimated in several speculations to be comparable to that of Earth's atmosphere at the top of Mount Everest.

The initial scientific reconnaissance of Mars in the 1960s by the United States and the Soviet Union

---

*The author, than a reporter on the *Indianapolis Times*, was called back to the newsroom that evening to work the story, which was slugged "Panic."

†Based on *Das Marsprojekt*, a special issue of *Weltraumfahrt* magazine, 1952.

dispelled the fictional image of Mars. This was replaced by a serial vision of a lunarlike, heavily cratered body in a sequence of U.S. *Mariner* fly-by missions in 1965–1969. *Mariner 4* flew by the southern hemisphere July 14, 1965. It returned 22 photographs that depicted a barren, icy wasteland. Flying by the planet in July and August 1969, *Mariners 6* and *7* sent back a less dismal image. Volcanoes and lava plains appeared. The northern hemisphere looked more Earthlike. Radiometric soundings returned evidence of water ice in the polar caps, the most conspicuous features on Mars.

It was not until the first Mars orbiter, *Mariner 9*, circled the planet in 1971 that the dual physiography of the hemispheres was apparent. In contrast to the ancient cratered regions of the south, the northern hemisphere appeared more varied, with its giant volcanoes, lava plains and equatorial canyonlands. North of the equator rose chains of volcanoes that dwarfed the largest on Earth. Great cracks and fractures could be seen in the crust. Mariner 9 photographed one of the remarkable features of the Solar System—a canyon 3000 mi long, 43 miles wide, and 4 miles deep. This grand canyon was named Valles Marineris. More big volcanoes appeared near the equator.

Along the equator ran chains of mountains. In this region the photos showed sinuous, rill-like features running hundreds of miles. They looked like ancient river valleys. Into them merged networks of smaller valleys that suggested dried up tributaries. The scenes from orbit suggested aerial views of dry streambeds during a drouth in the American Southwest. The "canali" were real after all, but did not resemble the features reported by Schiaparelli and other astronomers. Those were optical illusions created by autumn dust storms, such as the one that blanketed the surface when *Mariner 9* arrived in November 1971.

The physical characteristics of Mars were known in broad outline. Its resemblance to Earth was striking, considering the diversity of planets in the Solar System. Its equatorial radius is 3390 kilometers (2106.4 miles), a bit over one half that of Earth and nearly twice that of the Moon. The axis is inclined 25 degrees to the ecliptic plan (compared with Earth's axial inclination of 23.5 degrees) so that Mars has four seasons, like Earth's. The Martian seasons are not symmetrical because of the high eccentricity of Mars' orbit. As will be shown later, summer in the southern hemisphere is shorter and hotter than summer in the northern hemisphere.

Among proponents of Mars expeditions, the

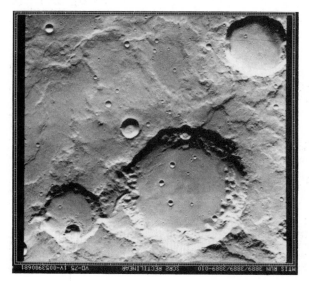 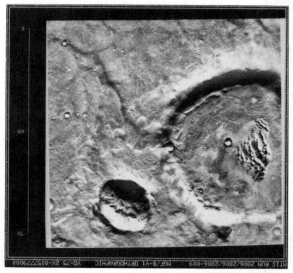

Figure 8.1. Craters on Mars vary by latitude. The crater at the left displays "creep deformation" because the ground has been softened by ice. It is located above 30 degrees north latitude. The crater at the right, located below 30 degrees north, does not show ground softening. (NASA-Ames)

moons of Mars have attracted increasing attention. These meteoroidlike bodies appear to be carbonaceous chondrites, containing water and organic compounds. Phobos, the inner moon, has been proposed as a potential base for the exploration of Mars.

The atmosphere of Mars is 95.3 percent carbon dioxide, 2.7 percent nitrogen and 1.6 percent argon, with smaller amounts of oxygen, carbon monoxide, and water. Surface temperature varies daily, geographically, and seasonally, as on Earth. There the resemblance ends. The Martian atmosphere is so thin (0.7 percent, Earth sea level pressure) that explorers would have to wear space suits and carry oxygen, as on the Moon. As on the Moon, habitats, laboratories, and workshops would have to be pressurized and shielded from cosmic radiation.

Mars is cold. During the summer of 1976 at its

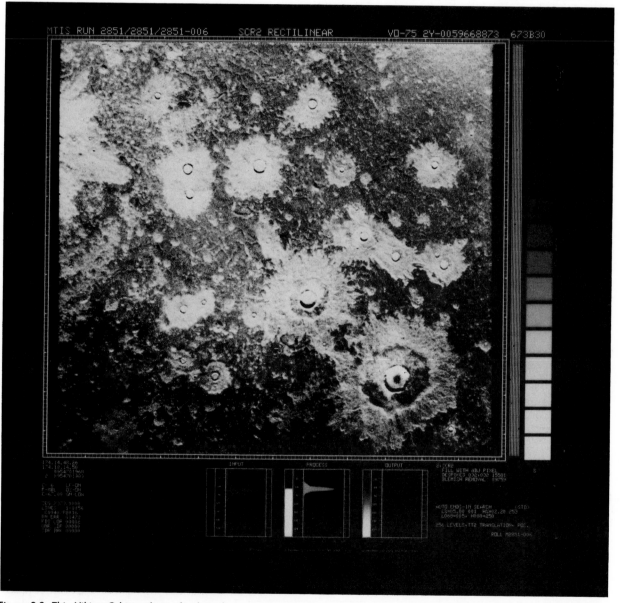

Figure 8.2. This Viking Orbiter photo of a heavily cratered surface in the northern hemisphere of Mars reveals a mottled landscape about 190 miles in area. The mottled appearance is due to the contrast between a dark, relatively smooth surface and brighter material excavated by impact. (National Space Science Data Center and NASA)

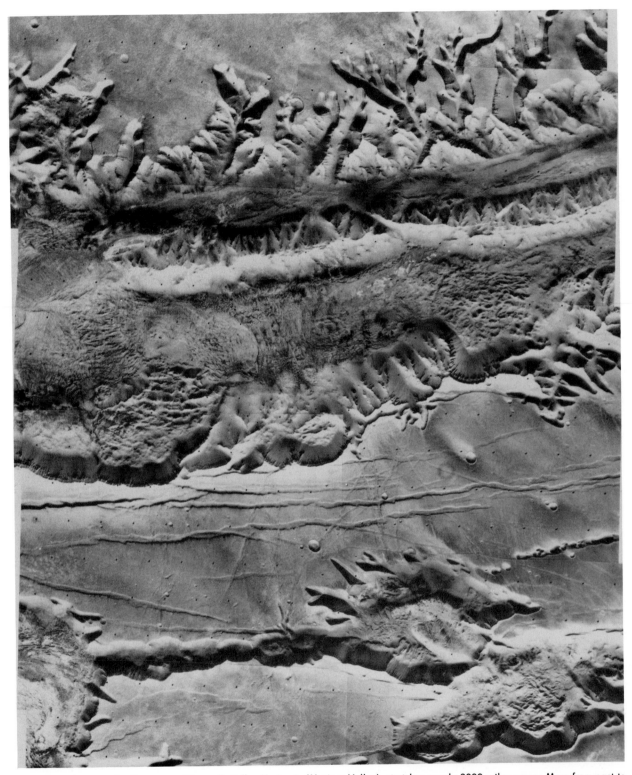

Figure 8.3. The "grandest" canyon of them all, Valles Marineris (Mariner Valley), stretches nearly 3000 miles across Mars from east to west. The main canyon crossing the lower part of this photomosaic has an apparent width of 43 miles and depth of 4 miles. It is part of a complex of canyon lands ranging more than 400 miles from north to south across the central disk of the planet. (NASA)

subtropical site, *Viking 1* lander reported temperatures of −85° C (−121° F) at dawn to −29°C (−20.2° F) at midafternoon. In winter, temperature at the southern ice cap was estimated at −133° C (−207.4° F). Fierce dust storms rage in the thin atmosphere, arising at midsummer in the south to sweep over most of the planet.

In 1971 *Mariner 9* while orbiting Mars was prevented from photographing surface features for three months by a dust storm. Such storms are periodic. Most of the dust eventually settles, but a residue remains to color the sky pink.

VIKING

Viking, launched during the opposition of 1975 (when Earth and Mars were aligned on the same side of the sun), was the most complex planetary mission of the space age. Logistically and scientifically, it was highly successful, although the results of its primary task—the search for evidence of life —were ambiguous. The Viking answer to the question, "Did life exist or had it ever existed on Mars?" was neither positive nor negative. It was not proved.

This ambitious and sophisticated expedition dispatched two dual spacecraft, *Viking I* and *Viking II*. Each vehicle consisted of an orbiter and a lander. The orbiter, with a mass of 5124.9 pounds, was fabricated from Mariner technology. The lander, with a mass of 2632.9 pounds was developed from lunar Surveyor technology, but it was much more advanced. Orbiter and lander with a combined mass of 7757.8 pounds were launched as a single spacecraft by a Titan 3–Centaur, the most powerful booster in the American inventory next to the Saturn 5 at that time. After each dual spacecraft encountered Mars, the lander separated from the orbiter and descended to the surface by means of parachute and retrorockets. The landings were targeted for northern hemisphere sites reconnoitered as level and landable by orbiter imagery.

Viking I was launched August 20, 1975 and was followed by the launch of *Viking II* September 9, 1975. The *Viking I* lander came down in a rock-strewn desert euphemistically called Chryse Plani-

Figure 8.4. The dark, circular features on the illuminated area of Mars are giant volcanoes, clearly discernible to the cameras of the approaching Viking I orbiter from a distance of 348,000 miles. The row of three volancoes in the upper center are the Tharsis Mountains, rising 12½ miles above the plains. Visible at the top is Mount Olympus. The light-colored circle at the bottom of the picture is the Argyre Impact Basin, apparently covered with light-reflecting frost. (NASA)

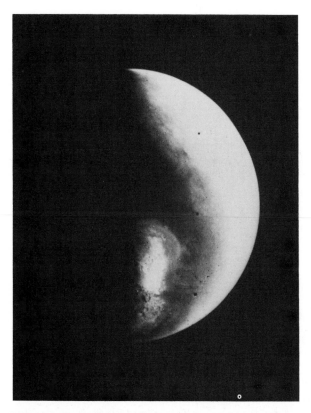

Figure 8.5. The half disk of Mars as it appeared to the Viking I Orbiter telescopic camera June 16, 1976 as the spacecraft approached the planet. In the bright area (frost) near the bottom of the picture is a circular impact basin, Hellas. Viking I was 425,000 miles from Mars. (NASA)

tia (the Golden Plains) at 22.4 degrees north latitude and 48 degrees west longitude July 20, 1976. It was a summer afternoon there, the seventh anniversary of the first manned landing on the Moon. The air temperature was 31 degrees below zero Fahrenheit (238 K).[2]

The first color photograph received from the lander at the Jet Propulsion Laboratory in Pasadena showed a rusty, rock-littered upland strikingly similar to the Arizona desert scape along Interstate 10, but without the utility poles. Viewers were astonished to see a light blue sky in this picture. But the color was false and when color values were corrected by the imaging team the sky became a light pink. It was the color of the red dust in the atmosphere.

The *Viking II* lander came down September 3, 1976, in a rolling rockscape called Utopia Planitia (Plains of Utopia) at 44 degrees north latitude and 226 degrees west longitude. It was mid-morning in late summer there at the landfall. The landscape depicted by the lander camera was similar to that viewed by lander I. It was colder there, 52 degrees below zero F.

At each site, the density of the carbon dioxide atmosphere was slightly higher than 7 millibars, that is, .7 percent of sea-level atmospheric pressure on Earth.*

Each lander carried an automated chemical laboratory, two computer centers, a television studio, a weather station, a seismometer, and dual power stations, one a photovoltaic system for converting sunshine into electricity and the other, two radioisotope (plutonium 238) thermoelectric generators. Each of the RTGs produced 70 watts.

The laboratory contained three separate incubators for the detection of biological activity and growth by microorganisms in the soil. The soil was collected by an automated backhoe and a scoop on an arm atop the lander. The scoop dumped gram lots of soil into conveyors that distributed it to the incubators. Analytical procedures for detecting microbes in the soil were analogous to those that would be done in a fully equipped laboratory on Earth.

---

* Sea-level atmospheric pressure on Earth is 1013 millibars or 1 bar.

Standing 5 feet above ground, the lander cameras—there were two—provided a view that a person holding a camera would see. The cameras could be operated independently, to photograph panoramas encircling the lander, or together, to transmit stereo views and give an impression of depth to the landing site.

Most of the photos were black and white but a color system in each camera transmitted data showing hues of surface and sky. The color signals could be calibrated to match the colors of the American flag painted on the side of lander. The impression of an American southwestern desert these photos transmitted lent an aura of familiar reality to the Martian scene, despite the alien sky. The Martian plains exhibited rocks of all sizes and shapes, mostly light gray to nearly black in color, but with a reddish tint. Some showed bubbles of gas, evidence of their volcanic origin. Their varied shapes suggested that they had been thrown out of erupting volcanoes.

On the Moon, surface rocks were fragments or amalgams of older rocks. They had been broken and in some instances recompacted by meteorite impacts. But here on the plains of Mars the rocks were clearly the residue of volcanic eruptions. The extent of Martian volcanism had not been perceived until *Mariner 9* made its orbital reconnaissance in 1971. The spacecraft had been blinded by a dust storm that obscured the surface. As the storm waned and the dust settled, the rim of a huge crater identified earlier as Nix Olympica (Snows of Olympus) materialized as a massive volcano 20 degrees north of the equator. It was greater in extent than any volcanic feature on Earth, and the name was promptly amended to Olympus Mons. It is the largest volcano yet seen in the Solar System.

Mount Olympus on Mars has been characterized as a shield volcano, its steep slopes built by eons of successive lava flows. *Mariner 9* and *Viking Orbiter* photos depict it as a distinguishing feature of Mars, like the crater Copernicus on the

Figure 8.6. The automated scoop on Viking Lander I is extended preparatory to collecting a soil sample for analysis in the Lander's biochemical laboratory. A box at the lower right housed the meteorological instrument which is deployed forward. This photo was transmitted August 20, 1976. (NASA-JPL)

Moon. Olympus is 372 mi across and 14 to 18 mi high. It is ringed by arcuate ridges and troughs for more than 300 mi beyond the lower slopes.

About 700 miles to the southeast, nearly on the equator, stood a row of three great volcanoes forming the bulk of the Tharsis Ridge. They were named Arsia Mons, Pavonis Mons, and Ascraeus Mons and were spaced about 435 miles apart on a fracture zone running northeast and southwest. Their

caldera summits rose 16.7 miles above datum (the mean elevation of the planetary surface) and their shield diameters ranged from 217 to 248 miles. By Earth standards, they were gigantic. Arsia Mons exhibited a caldera (crater) 68 mi across, compared with the 1.6 mile-diameter caldera of Mauna Loa in Hawaii. The Ascraeus caldera was 2.2 miles deep compared with the 656 foot depths of Mauna Loa's crater. Alba Patera at the northern edge of the

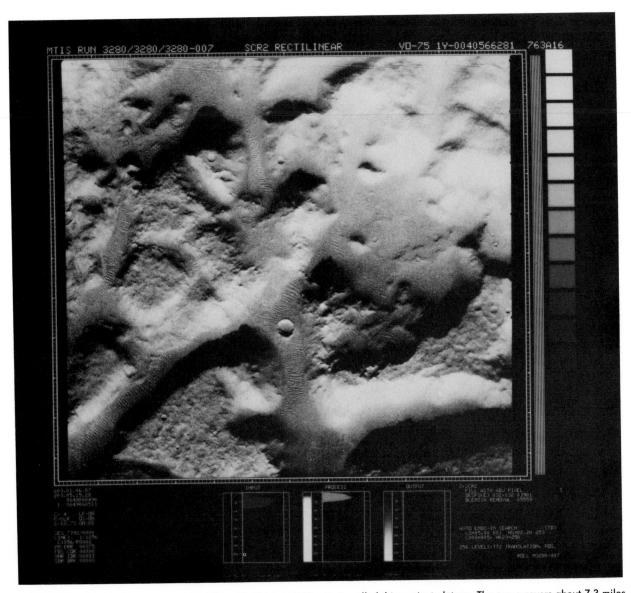

Figure 8.7. An "engimatic landscape" the Viking photo interpretation team called this ancient plateau. The scene covers about 7.3 miles of a surface cut by valleys in which wind blown debris has accumulated. Chains of low sand dunes can be seen running transverse to the valleys, especially near the crater at the upper center. (National Space Science Data Center and NASA).

Tharsis Ridge covers the largest area of any of the ridge volcanoes and has a caldera 372.8 miles in diameter, but its summit is only about three miles. In its report on the "Geology of the Terrestrial Planets," edited by Michael H. Carr, NASA noted that although the type and location of Earth's volcanoes are determined mainly by plate tectonic activity, this is not the case on Mars, where plate structure does not appear to exist. Other large volcanic piles had been photographed by *Mariner 9* in the region called Elysium. The largest, Elysium Mons, was 155 mi across and 9.3 miles high. Lesser volcanoes were observed in the southern hemisphere.

It was reasonable to suppose that from these massive vents in the martian crust, an atmosphere and hydrosphere had been vented from the molten interior, as on Earth. What had become of them? The 7 millibar atmosphere *Viking* measured on the plains was far too thin to hold liquid water on the surface. Yet the sinuous rills, first seen in *Mariner 9* imagery, resembled terrestrial river valleys, and connecting threads resembled tributary streambeds. There were depressions shaped like dry lake beds, some large enough to suggest inland seas.

From this remarkable visual evidence, observers could assume the existence of an ancient hydrosphere, although the assumption could be challenged by the argument that these features could have been created by lava flows. The predominant view, perhaps supported by preference, held that it had rained on Mars in an epoch when the atmosphere was dense enough and the temperature high enough for the existence of surface water. The view supported the rationale for a renewed search for life, which Viking had not found and which biologists were sure could not have originated without free water.

The assumption of an ancient hydrosphere on Mars implied a climatic change to account for the disappearance of rivers, lakes, and seas and the dissipation of a denser atmosphere. Several theories of climatic variation had been proposed. The most obvious was that Mars is presently in an ice age in which much of the atmosphere and hydrosphere lie frozen at the poles and in the ground. Inasmuch as Earth has experienced repeated episodes of glaciation, the idea that a more temperate environment existed on Mars in the past has support. Climatic changes on both planets would appear to be driven by changing relationships of portions of their surfaces to the sun and by the

Figure 8.8. A thin coating of water ice is visible at the Lander 2 site of Viking at Chryse Utopia, 42 degrees north latitude. The ice, about 0.002 centimeters thick, forms a thin film on the shaded side of the rocks. It consists of water and carbon dioxide condensed on dust particles. (National Space Science Data Center and NASA).

eccentricity of their orbits. Whether these changes have been severe enough to have depressurized Mars and plunged it into a glaciated state is speculative, and requires further investigation.

Air pressure on Mars varies by season even now, however, because a significant fraction of the carbon dioxide atmosphere condenses upon the winter polar cap. That is, part of the atmosphere is precipitated as carbon dioxide snow. Because the southern winters are longer and summers shorter than the northern ones, the southern ice cap is more extensive at maximum and smaller at minimum than the northern cap.* For that reason the southern ice cap may control seasonal changes in atmospheric density of Mars.

At Chryse Planitia, the site of the *Viking* I lander, atmospheric pressure ranged from 6.9 millibars at

the time of the southern midwinter to 9 millibars at the end of the southern spring. At that point, the southern cap reached its minimum extent and the northern cap had not yet expanded greatly. The site of lander I was about 1.24 miles below Mars datum. At the site of lander II, about 4000 miles to the north in the subpolar region of Utopia Planitia, the pressure ranged from 7.5 to 10 millibars. The site was 1.86 mi below datum, in a rocky, bowl-shaped depression.[3] At peak pressure in the northern spring, the atmosphere reaches 1 percent of the pressure of Earth's atmosphere at sea level.

IN COLD STORAGE

Advocates of the manned exploration and settlement of Mars express certainty that there is plenty of water on Mars, in the polar glaciers, permafrost,

---

* A result of the eccentric orbit of Mars.

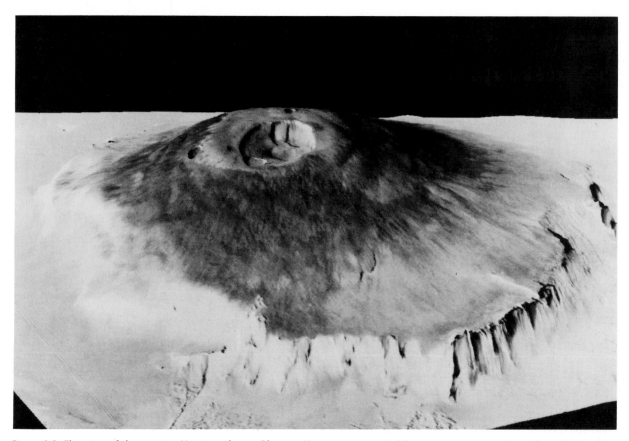

Figure 8.9. This view of the massive Martian volcano, Olympus Mons, was generated by computer, as seen from the north. It extends 370 miles across and rises 16 miles above the surrounding plains. These dimensions make it the largest volcano yet discovered in the Solar System. It is a shield volcano similar in structure but much larger than the Hawaiian volcanoes. (U.S. Geological Survey)

Figure 8.10. The Viking I lander site at Chryse Planitia appears to be a quiet place where nothing seems to happen. But these photos taken by the lander camera show that something does. The upper photo was taken in August 1976. When the scene was photographed again about a year later, there were two changes. They are marked in the lower photo. At "A" a dust slide has occurred at the right end of a structure called "Whole Rock" at the upper right. At "B", a dust slide has exposed part of a buried rock beside "Big Joe" boulder in left center. (National Space Science Data Center and NASA)

and adsorbed in underground soil. A corollary has it that a large part of the atmosphere also is in cold storage in the winter ice caps. If evaporated, it would raise pressure sufficiently to support liquid water on the surface.

Most of Mars' volatile elements and compounds that have outgassed from the interior are trapped within the body of the planet, in the view of Michael H. Carr, U.S. Geological Survey. Basalt and clay have the capacity to adsorb carbon dioxide. Carr and his colleagues suggested that the crust could have adsorbed carbon dioxide to a depth of 6 miles. The surface may contain significant amounts of water as ice.[4]

Carr estimated that about 100 m of water (that is, a volume that would cover the whole planet to that depth) was outgassed. About a third of it is tied up in the polar caps, in layers of dust and soil laying beneath or around the perimeter of the ice caps or as underground permafrost or water chemically bound to the subsoil. The remainder may exist as groundwater or ice miles beneath the surface of the Martian crust.

Wet and Dry Ice. On Mars the south polar ice cap in winter extends northward about 50 degrees of latitude. The northern ice sheet reaches 30 degrees of latitude in winter. For comparison, the farthest extent of Earth's antarctic ice sheet is just about 23.5 degrees; that is, to the antarctic circle at 66.5 degrees south latitude.

On September 30, 1976, the *Viking II* orbiter was shifted to a polar orbit. Looking down at the poles, its instruments detected a large variation in water vapor. It was winter at the south pole, where atmospheric water vapor was near zero, having precipitated as ice. As the orbiter flew northward, the amount of water vapor increased over temperate and equatorial zones and then decreased as the spacecraft approached the north polar region. At the north pole, where the surface temperature registered minus 96 F, atmospheric moisture was again near zero. However, the north polar temperature was still above the freezing point of carbon dioxide and the thin layer of carbon dioxide frost or dry ice, precipitated during winter, had sublimed and returned to the atmosphere. However, the cameras

still saw a substantial sheet of ice. It was clearly water ice and appeared to be permanent.

The cameras showed breaks in the north polar ice sheet. They appeared to be steep, ice-free valleys. Around the edge of the ice sheet and in ice-free gaps appeared layers of wind-deposited dust, forming a turbulent surface of hundreds of square miles. Mounds of dust several miles in depth were seen. Orbital photographs have shown that these and thick, layered deposits extend from each pole outward to 80 degrees of latitude. The deposits are believed to be a mixture of volatiles and dust, mostly covered with ice in winter, exposed in summer. The seasonal cap of dry ice that forms around each pole in autumn dissipates in spring. It may be thin, only a few tens of feet thick. At maximum, the south polar ice reaches 40 degrees south latitude and the north polar ice extends to 60 degrees north latitude.

As the dry ice disappears from the south in spring, it leaves exposed the permanent cap of water ice, which is estimated to be about 220 miles across. The center of the cap is offset 4 degrees from the pole along 30 degrees west longitude. In the north, the summer residual cap of water ice is about 600 mi across, about the same in extent as the layered dust/ice deposits. Although the summers are longer in the north, temperatures do not now rise high enough to melt the water ice, and it appears to be permanent in the present epoch.

The layered deposits of dust are interpreted by some observers as evidence of a denser atmosphere in the past. The present atmosphere appears to be too thin to deposit such a volume of dust as appears now. The layering of the deposits of dust implies not only that past climate was different but that it was subject to periodic variations.

Still, even the present thin atmosphere moves enormous volumes of dust during periodic dust storms that obscure the planet surface. With the onset of the first dust storm in 1977, average wind speeds of 6.5 to 22.9 feet per second increased to 55.7 feet per sec, with gusts up to 98.4 feet per second, Viking landers reported. The dust storms of 1977 appeared to start along the edge of the south polar cap.

THE SUNSHINE FACTOR

An explanation for the climatic changes that physical evidence suggests as having accounted for the present environment of Mars is the hypothesis that the intensity of sunshine received at Mars has varied over periods of a million to 100 million years. Moreover, sunshine falling on polar regions has varied specifically with changes of the tilt of the axis of rotation to the plane of the orbit.[5] Periodic changes in axial tilt, or obliquity, have occurred on Earth and are cited as one of several causes of repeated ice ages.

William R. Ward of the Jet Propulsion Laboratory suggested that the tilt of the Mars axis may have been altered by the pull of the sun over time. Presently, the tilt is 23.9 degrees from the orbital plane, compared with Earth's tilt of 23.5 degrees, so that Mars and Earth have similar seasons. The axial tilt of Mars has been calculated to vary from 15 to 35 degrees over a period of 100,000 to a million years. Recent calculations indicate that the tilt was as much as 45 degrees billions of years ago, before the rise of the Tharsis volcanoes.[6]

Such variations in tilt would be expected to affect the angle at which sunshine would strike polar and other regions of the planet. On Mars, as on Earth, the poles are cold because sunshine strikes them obliquely during the half year when these regions are fully illuminated and barely at all during the half year when they are not. The illumination of the polar regions may have been more intense in the past, inhibiting atmospheric precipitation in winter and preserving a denser atmosphere over the planet.

It has been known that some of the atmosphere is stored at the winter pole since *Mariner 7* data showed that south polar temperatures were low enough to precipitate part of the carbon dioxide atmosphere in winter. That suggested that if the ice caps were composed entirely of carbon dioxide, a rise in the present temperature regime at the poles would vaporize it and raise atmospheric pressure planetwide to about 30 millibars, or about 3 percent of Earth's sea-level pressure. A rise in temperature would occur if the planet's axial tilt increased to 35 degrees. This would result in an increase in the intensity of sunshine reaching the polar regions, according to calculations by Bruce C. Murray and colleagues at the California Institute of Technology.[7]

Following the temperature data from *Mariner 7* (1969) suggesting that some of the atmosphere was stored as ice at the poles, evidence that carbon dioxide was precipitated out as ice at the winter pole was seen in the data returned by *Viking* in 1976–1977. During the first six months the two *Viking* landers were on Mars, their weather stations reported a 5 percent drop in atmospheric pressure with the arrival of winter in the northern hemisphere.

Having observed the process by which carbon dioxide changed its state at the poles, observers were moved to wonder why the caps did not disappear altogether at the summer pole. During the summer of 1976, the *Viking II* orbiter radiometer reported north polar temperature at 205 degrees Kelvin (minus 90.4 F.) At 6 millibars average pressure, the temperature was too high to maintain a permanent ice cap, but photos showed that one was there.

Scientists studying the Martian atmosphere then concluded that the permanent ice at both poles is water ice, overlaid seasonally by the crust of dry ice.

In the light of recent studies, Carr has estimated that enough water was ejected from the interior of Mars by early volcanism to have covered the surface to a depth of three to six tenths of a mile. Outgassed with it was the equivalent of 10–20 bars (Earth atmospheres) of carbon dioxide and 0.1 to 0.3 bars of nitrogen.[8] Carr opined that the flow of the ejecta on Mars was more fluid than on the Moon or Mercury and extended for greater distances from the rim crests of the volcanoes because of the inclusion of ice, water, and atmosphere in the debris. The geologist conceded that his estimates are higher than those made earlier but are consistent with the surface geology. Much of the primordial water is still within the body of planet, soaked up by pulverized rocks smashed by the great meteorite bombardment of 3.8 billion years ago that hit all the terrestrial planets, he said. Much of the outgassed carbon dioxide may have been adsorbed by the water.[9] Future investigation may find a layer of carbonate rock 200 to 500 meters (656–1600 feet) thick. Now widely held, the belief that underground water resources can be tapped by deep drilling makes this effort a major objective of 21st century exploration. The planet's crust has substantial water-holding capacity, Carr said.

Many of the large outflow channels photographed on the planet may have been formed by the eruption of ground water under pressure be-

Figure 8.11. A network of channels interpreted by scientists as ancient river valleys is clearly visible in this orbital photo of the Martian landscape. In the center of the area, a dendritic pattern of stream beds appears to empty into a broad embayment. (NASA-Ames).

neath the permafrost, which Carr estimated was 0.62 miles thick. The volume of water flowing through the channels must have been at least as large as the volume of soil it eroded to form the channels, he said. The volume eroded from channels in the Chryse region indicated water volume equivalent of 5 million cubic kilometers (1.2 million cubic miles), enough to cover the entire planet to a depth of 115 feet.

The channels have been classified into three types—fretted, runoff, and outflow. Fretted channels are broad, flat, floored valleys with steep walls. Most of them have tributaries that become nar-

rower upstream. Fretted channel lands lie in two belts, 25 degrees of latitude wide, centered on 40° north and 45 south. Runoff channels start small, increase in size downstream, and have a V-chaped cross section. The common type is a simple gully.[10]

Some outflow channels have been formed by catastrophic floods. They emerge from chaotic terrain and are widest and deepest close to their sources. Many are large by terrestrial standards. One channel emerging from Juventae Chasma and extending northward for several hundred miles across Lunae Planum has scoured a swath more than 60 miles wide. Sections of the channel in

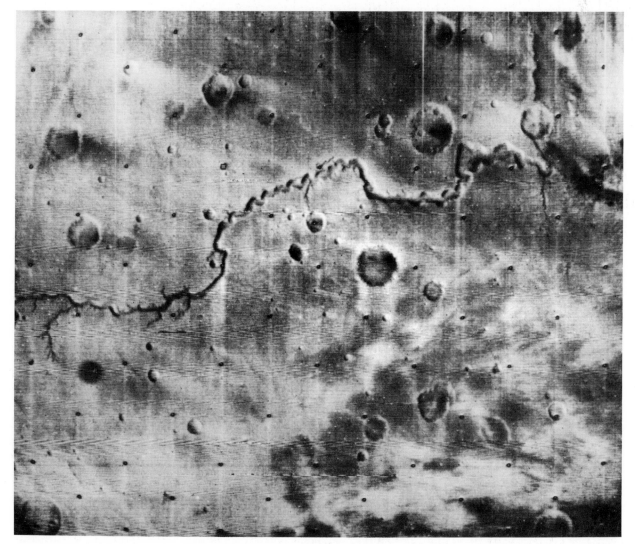

Figure 8.12. More evidence of ancient rivers on Mars appears in this Mariner 9 photo of a 355 mile long channel 3 to 3½ miles wide. (NASA-Ames).

Kasei Vallis are more than 120 miles wide. Ares Vallis, a single, discrete channel in the uplands south of Chryse Planitia, pools to a width of 186 miles and is 11 to 18 miles wide over most of its length. Characteristically, outflow channels have branches that diverge around obstacles such as low hills and craters, longitudinal grooves, teardrop-shaped islands, and arcuate escarpments. These were traced in photographs as running about 600 miles northward. Flow lines then turned to the northeast and converged in several channels cut in ancient, cratered terrain between the regions called Lunae Planum and Chryse Planitia. The flow lines run across the plains to form a complex pattern of broad, shallow channels, elongated islands, and longitudinal grooves.[11] The runoff channels appear to be mostly old and the outflow channels have a wide range of ages, some predating 3.9 billion years ago. Most of the large channels around the Chryse basin appear from crater densities to have been formed before 2.5 billion years ago.

The Martian outflow channels have been compared to the channeled scablands of eastern Washington, where the largest known terrestrial flood feature is located. The state's scablands were formed in the late Pleistocene Epoch, when an ice dam that had impounded the waters of Lake Missoula gave way. The lake was vast, covering large parts of Idaho and western Montana. Peak discharge of the lake water has been estimated at 10 million

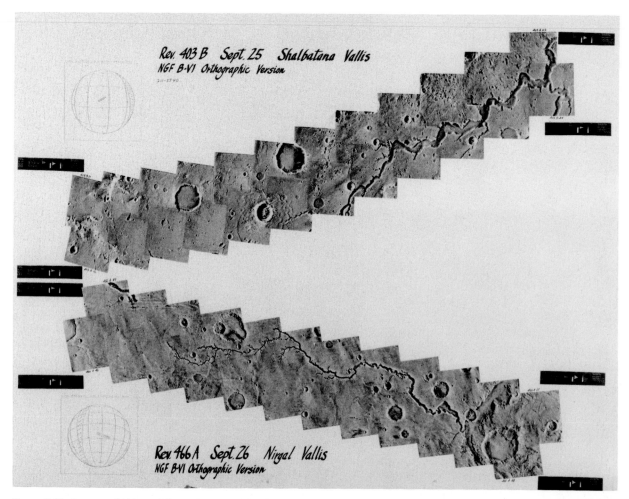

Figure 8.13. Sequential Viking Orbiter photos in 1976 show extensive river valleys in two areas, the Shalbatana and Nirgal Valleys. The Nirgal Vallis channel appears to have its origin in a deeply fissured region around a large crater. (National Space Science Data Center and NASA).

meters a second. The flood probably was caused by the melting of an ice sheet by volcanic heat.

Analysis of Martian surface features that look like lake beds and shallow seabeds indicate an outflow of 20 million cubic meters of water a second and channels 600 mi long, according to Timothy J. Parker and colleagues at the Jet Propulsion Laboratory.[12] Parker said that *Viking* imagery indicates depressions that could be interpreted as seabeds, but the features could not be resolved. Curvilinear ridges that could have been made by waves and strand lines marking high water were tentatively identified. But most of the channels emptied into featureless plains without evidence of sedimentary deposition or of deltas at the channel mouth, Parker and his colleagues said. "Apparently, catastrophic deposition spreads material over a much larger outwash area than that of a typical river," they added. "On Mars, this effect could be more pronounced due to the lower gravity."[13]

The evidence for the supposed rivers, lakes and seas was presented at a symposium on the evolution of the Martian climate and atmosphere July 17–19, 1986 in Washington, D.C. Without ground truth beyond the *Viking* landing sites, the analyses relied on narrow bands of surface imagery from orbit and inductive reasoning. Mars needed a Marco Polo.

Estimates of how much water there may have been on the ancient surface of Mars varied from Carr's calculation of a global layer up to 1 kilometer deep to that of Parker and his colleagues of a global layer only 10 meters deep. Some critics rejected water as the agent, arguing that these features

could have been carved by lava flows. Stephen Clifford of the Lunar and Planetary Institute, Houston, estimated that the regolith (surface rock) had the capacity to hold a volume of water equivalent to a global layer hundreds of feet deep. There was a bit of evidence. The operation of the Viking molecular analysis experiment extracted some water from the soil when samples were heated. The molecular analysis team said that water from a sample taken from under a rock at Utopia when the sample was heated may represent adsorbed water. If Mars is like Earth in this respect, adsorbed water may be present at larger depths, the team reported. It added, "Thus, the Martian regolith may contain substantial amounts of chemically bound and adsorbed water."[14]

AN ECCENTRIC PLANET

Beyond the inclination of its axis to the plane of its orbit, two other physical characteristics of Mars influence its climate and the density of its atmosphere.

The axis of rotation precesses, like a spinning top. It describes a circle among the stars in a period of 50,000 years. This means that climatic conditions are likely to vary over a cycle of 25,000 years as the relationship of the polar regions to the sun changes with precession

The other characteristic is the eccentricity of the orbit of Mars, the third most elliptical in the Solar System. Eccentricity of the Mars orbit is 0.0933 compared with 0.0167 for Earth.

Figure 8.14. The sinuous channel in this photo is not on Mars, but on Earth. It is the Ohio River at Parkersburg, West Virginia, imaged by synthetic aperture radar aboard NASA's experimental Seasat satellite August 26, 1978. (NASA)

Because of this difference and because the seasons on Mars are longer than the comparable ones on Earth (the Martian year is 686.98 days compared to Earth's 365.25), the effect of orbital eccentricity on the intensity of seasonal sunshine is thought to be greater on Mars than on Earth.

At perihelion, Mars is 26.45 million miles closer to the sun than at aphelion, while the difference between perihelion and aphelion in Earth's orbit is 3.22 million miles. Yet even this smaller difference is regarded by some climatologists as an influence on Earth's climate, especially when accompanied by a change in axial tilt. On Earth as on Mars, summer comes to the northern hemisphere at aphelion. Although less sunshine is received by the northern hemisphere then than at perihelion, the period of direct sunshine falling on the hemisphere is longer than at perihelion. It takes the Earth longer to pass around the more distant segment of the orbit at aphelion. Earth's spring–summer is seven days longer in the north than in the south.

Mars experiences an analogous effect. Its north-

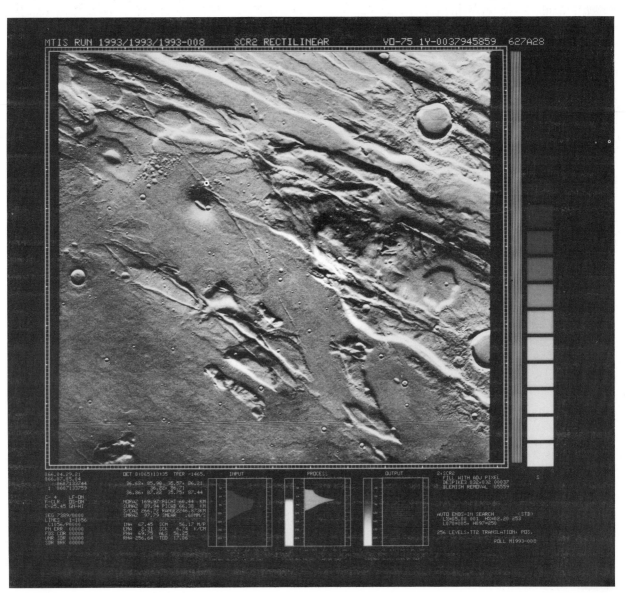

Figure 8.15. Apparent broad outflow channels and embayments on Mars reveal a history of torrential outflows of ground water in a past climatic era. (National Space Science Data Center and NASA)

ern summer occurs at aphelion too, but at Mars the effect is more pronounced. Because Mars is moving more slowly in its orbit at aphelion, the spring–summer season in the northern hemisphere (382 days) lasts about 76 days longer than it does in the southern hemisphere (306 days). The northern hemisphere fall–winter season is correspondingly shorter. However, because Mars is 26.4 million miles farther from the sun during its northern hemi-

sphere spring–summer season, the planet receives 44 percent less radiation at aphelion than at perihelion.[15] Depite the longer spring–summer, the northern ice cap, reduced to water ice in summer, never disappears. In the south the summers are hotter but shorter (160 days compared with 182 days in the north). The dry ice of the southern cap vaporizes. When viewed from Earth, the ice cap becomes hidden by a hood of vapor, but the water

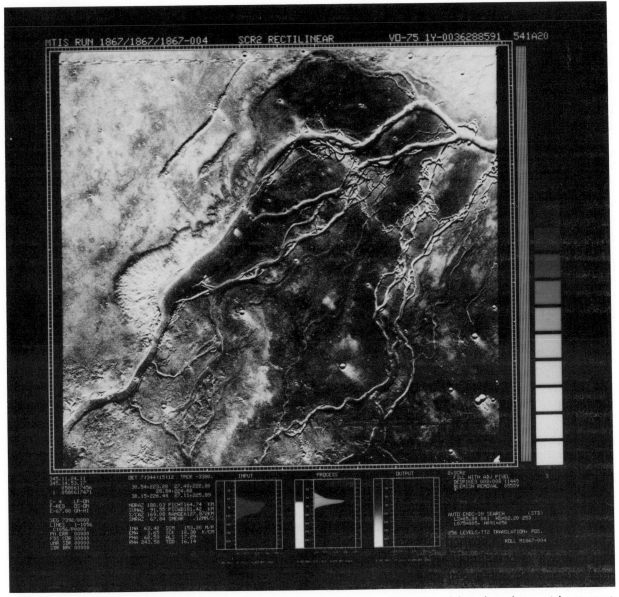

Figure 8.16. Mars' heavily channeled landscape extends through equatorial regions. The wide and deep channel upper right appears to be older than the crater in its center.

ice remains. The temperature never rises enough to melt it.

Analysis of *Mariner 9* photos and radiometric data indicated that a reservoir of carbon dioxide 300 to 1000 meters thick lies under the permanent glacier of water ice at the north pole.[16] If changes in obliquity and precession reversed the seasons on Mars, bringing summer to the northern hemisphere at perihelion, a climatic change might be expected. Even though shorter, a hotter summer at the north pole could release water vapor and carbon dioxide to the atmosphere in sufficient volume to warm the climate and raise atmospheric density to the point where liquid water could exist on the surface. Such a reversal in seasonal insolation may account for episodes of water flow. Whatever the cause, the evidence that streams and rivers have flowed on Mars is highly visible.

THE WEATHER AND SOIL

The cameras of *Viking I* showed close-up views of the soil around the lander's three legs. Where the landing feet rested, the soil appeared to have the consistency and pattern of caliche, a stony soil in arid lands of the American southwest. At Chryse Planitia, the soil showed fractures in a polygonal pattern called *duricrust*. On Earth, duricrust is formed when a dilute solution of salts migrates up through the soil and evaporates. Viking scientists suggested that the same process may have occurred on Mars.[17] Soil particles clinging to magnets on the legs of the landers showed the presence of magnetite, an oxide of iron. In the middle distance of the camera's view appeared duney undulations composed of clumpy soil and rock. It was colored orange by iron oxides. The skies were clear during the day, although thin, light clouds and wisps of fog appeared in the mornings.

By contrast, *Mariner 9* had gone into orbit during a planet-wide storm. Although dust storms had been seen before by telescopic observation from Earth, their magnitude and opacity were surprising. An orbiting satellite was blind during such a storm until the dust settled.

In the spring of 1977, when the *Viking* landers had been on the surface for nearly a year, dust clouds rose out of the southern hemisphere and a new storm began. The storm seemed to originate in the southern hemisphere as Mars approached perihelion in its eccentric orbit. This time the *Viking* orbiters were there to photograph the evolution of the storm and its subsidence. But the atmospheric temperature processes that generated such violent activity in such a thin atmosphere remained to be identified.* When the dust settled, the weather turned cold and clear—as usual. The lander weather stations consistently reported light winds from the east in late afternoon changing to light winds from the southeast after midnight. Maximum wind speed was 15 miles per hour. Temperatures at both sites, which were 4000 miles apart were nearly the same, ranging on the average from -122 F. after dawn to -22 F. in mid-afternoon.

A day on Mars (24 hours 37 minutes) is only slightly longer than a day on Earth (23 hours 56 minutes), but for human observers on the planet the day–night cycle would be similar and the weather monotonous. During extravehicular activity (EVA), explorers would wear pressure suits and carry oxygen as they did on the Moon. They would find it necessary to adapt to a gravitational field only two fifths (.38) that of Earth.

For the geologist, the chemist, the climatologist, Mars would be fascinating. For the nonspecialist, it would certainly be a harsh, cold desert—interesting to visit, a barren place to live. Or so it appeared in the last quarter of the 20th century.

NITROGEN

The pattern of change in wind direction at Chryse Planitia was comparable to that of winds blowing across the American plains. There were occasional high clouds in the thin air. Although it holds hardly a thousandth as much water as Earth's atmosphere, enough is present to condense as fog. In the small valleys, moisture freezes at night and vaporizes at sunrise, condensing to patches of white fog. The fog dispels as the temperature rises.

During their descent the landers analyzed the

---

* In 1977, perihelion occurred in late April.

composition and pressure of the atmosphere in a vertical column. As mentioned earlier, the pressure at the landing sites was seven tenths of 1 per cent of Earth's atmospheric pressure at sea level. The composition was 95 per cent carbon dioxide, 2 to 3 percent nitrogen, 10. to 0.4 per cent oxygen and 1 to 2 per cent argon.

The finding of nitrogen, an essential element in protein, enhanced the possibility of finding living organisms on Mars. The small fraction of atmospheric oxygen was not construed as a negative sign. There was plenty of oxygen on Mars, bound up in the rocks and soil, as the red-rust color showed. It could be recovered as readily from rocks as on the Moon.

Viking chemical analysis revealed that the ratios of carbon and oxygen isotopes in the atmosphere are similar to the isotope ratios of those elements on Earth. The discovery was viewed by the science team as significant, because it demonstrated, as did the presence of water, that both Earth and Mars formed in the same region of the solar nebula. The ratios of carbon 13 to 12 and of oxygen 18 to 16 are identical in the atmospheres of both planets.[18] The science team found that nitrogen is an exception. The ratio of nitrogen 15 to 14 on Mars differs from that of Earth. The analysts supposed that the lighter isotope, nitrogen 14, has been lost to some extent with the loss of carbon dioxide from the atmosphere.

On the eighth day of its arrival July 20, 1976, the *Viking I* lander scoop picked up a soil sample for analysis. About 5 percent of the clump was seen clinging to lander leg magnets in a close-up view from the camera. It was magnetite (Fe_3O_4), an important industrial ore of iron. Its apparent abundance looked promising for future development.

The sample was analyzed by x-ray bombardment. Secondary radiation was diffracted by atoms in the soil at frequencies characteristic of certain elements. These and their approximate abundance were identified as silicon, 21 per cent; iron, 13; aluminum, 3; magnesium, 5; calcium, 4; sulfur, 3; chlorine, 0.7; titanium, 0.5 and potassium, less than 0.25. Oxygen was not detectable by this method, but it was estimated as 42 percent of the soil and rock. About 8 percent was believed to be sodium

and hydrogen, also not detectable by x-ray diffraction.

Soil composition was the same at both landing sites a continental distance apart. It corresponded generally to the composition of terrestrial or lunar basaltic lava, but with some differences. The Mars soil contained less aluminum than Earth basalt and less titanium than lunar basalt in these examples. The differences may be regional rather than planet-wide.

The relatively large amount of iron (13 percent) confirmed the pre-*Viking* assumption that the planet's red color is derived from iron oxides or rust. Iron is present in two compounds, magnetite and nonmagnetic oxide.

Unlike the Moon, the Mars soil contained about 1 percent water. The soil was a mix of original basalt with other compounds formed by atmospheric weathering. This process produced an iron-rich clay with other compounds, including iron hydroxide and magnesium sulfate.

LIFE

The question of life on Mars dominated the Viking project. It was also a central research objective of Soviet scientists, but their Mars exploration missions had been characterized by hard luck. The U.S.S.R. launched two Mars reconnaissance vehicles in 1960. Neither got out of Earth orbit. A third also failed in 1962. A fourth, announced November 3, 1962, as *Mars I,* flew by Mars, but its communications failed. *Mars II* consisting of an orbiter and a lander, went into Mars orbit November 27, 1971, 14 days after *Mariner 9.* The *Mars II* lander was released to descend to the surface; nothing more was heard from it. The back-up vehicle, *Mars III,* entered orbit in December 1971. Its lander, which the Russians said was equipped also to detect life, carried a mass spectrometer to identify the composition of the atmosphere and a television camera. About 20 seconds after the camera was turned on, communications failed.

Two more vehicles, *Mars IV* and *V* were launched in July 1973. *Mars IV* failed to enter orbit. *Mars V* entered Mars orbit February 12, 1974, and returned surface photos. *Mars VI* and *VII* were both launched in August 1973, carrying landers. The *Mars VII* lander attempt failed, but the *Mars VI* lander reached the surface March 12, 1974, and again communications went out.

During the *Mars VI* lander descent before communications failed, telemetry from the mass spectrometer atmosphere pump showed a high percentage of inert gas, thought to be argon. This report excited some members of the Viking team because it hinted at the existence of a denser atmosphere in the past. A relatively high concentration of argon in the Mars atmosphere could be interpreted as the residue of an ancient atmosphere of much higher pressure. If Viking, due to be launched the following year, confirmed this, the likelihood that life had started on Mars was vastly improved. Some members of the American scientific community were openly skeptical about this interpretation of the Russian data and the data's validity. It was not confirmed by Viking. Still, the biology experiments were the main event of the Viking program. They were devised on the assumption that living organisms would be based on carbon and would reveal their existence by transforming carbon compounds in the presence of water and by multiplying under benign environmental conditions of the biology laboratory.

It was assumed also that the micro forms of life, such as bacteria, would be the most abundant and the most likely to survive the changing Martian environment as evidenced by the dry river and stream beds. The experiment chambers were small incubators that would warm and nourish microbes and monitor their biochemical activity.

A possibility that Mars microbes might be adapted to cold temperatures and would react adversely or not at all to artificial warmth was raised as an issue during the mission. Experiments were modeled in the context of Earth conditions. A part of the model was the environment of an antarctic dry valley with a temperature regime analogous to that of a summer midday on the Martian plains. It was reasonable to believe that if microorganisms in the antarctic soil would flourish in a *Viking*-type incubator, Martian organisms would do so. This was the clos-

est preflight tests had come to simulating the Martian environment.

On Mars, automated chemical laboratories in both landers functioned with remarkable precision, but they yielded ambiguous results. The experiments were performed at both landing sites, and data on which the biology report was based were accumulated over a period of eight months. One of the three biology tests was based on the transformation of carbon dioxide by plants into organic compounds. It was called the carbon assimilation, or pyrolytic release, experiment. Radioactive carbon monoxide and carbon dioxide were added to the test chamber containing 0.25 cubic centimeters of soil. The sample was illuminated by a xenon arc lamp that simulated sunlight at Mars. After the soil was incubated for 120 hours at temperatures of 46.4° to 80.6° F., the radioactive atmosphere was flushed out of the incubator by an injection of helium.

The soil was then heated to 1184° F to vaporize any organic compounds synthesized by bacteria. The vapor was then oxidized by cupric oxide to form carbon dioxide again. If organic matter had been synthesized from the original radioactive gases, the radioactivity would reappear in the reformed gas. That is, if metabolism had taken place through photosynthesis, the metabolized carbon 14 in the reformed carbon dioxide would show it when the gas was passed through a radiation counter. Ten runs of this experiment showed that carbon assimilation took place, but it occurred also, although at a reduced, level when the sample was sterilized by heat at the outset. What was going on? There was no answer.

The Viking biology team concluded, "The data show that a fixation of atmospheric carbon occurs in the surface material of Mars under conditions approximating the surface ones. . . . This activity is quite small by terrestrial standards, but it is significant. . . . Nevertheless, a biological interpretation of the results is unlikely in view of the thermostability of the reaction."[19]

The leader of the biology team, Harold Klein of the Ames Research Center, told a new conference in Washington November 9, 1976, "These are the facts . . . they do not rigorously prove the presence of life on Mars. They do not rigorously exclude the presence of life on Mars." The only alternative explanation of the carbon assimilation that the experiment suggested was some undetermined, inorganic chemical process. As of this writing, it has not been identified.

A second experiment also yielded ambiguous but tantalizing results—the gas exchange test. A cubic centimeter of soil was dumped into a test cell by the automated scoop. A mixture of gases was added consisting of 91.65 percent helium, 5.51 percent krypton, and 2.84 percent carbon dioxide. A nutrient consisting of organic compounds and inorganic salts was injected into the cell. This was done first to humidify the soil and later to wet the sample. The initial reaction was surprisingly fast, starting with the humidification. Oxygen was released immediately, but instead of increasing in volume as expected, the release tapered off after only two hours. When additional nutrient was added to make the soil wet, additional oxygen failed to appear. However, carbon dioxide appeared until the eleventh day of the test.

The experimenters sought to determine whether the initial generation of oxygen was a biochemical product or the result of an inorganic chemical reaction. They heated the soil to a temperature of 293° F before repeating the test. This did not prevent the initial production of oxygen and indicated that the oxygen release was not biogenic. The effect was attributed to the presence of "superoxides" in the soil, formed by ultraviolet radiation. Because Mars has no ozone layer, the surface apparently is exposed to solar ultraviolet at possibly lethal intensity.

The strongest biological signal was interpreted from the results of the third test, the labelled release experiment. It was performed at both landing sites over periods of up to 60 days.

One half cubic centimeter of soil was deposited in the test cell filled with local atmosphere. The soil was moistened with 0.115 milliliters of nutrient containing formate, lactate, glycolate, glycine, and alanine and labelled with radioactive carbon 14, as a tracer. If this "soup" was ingested by living or-

ganisms, it was assumed that they would metabolize it and excrete labeled carbon monoxide, carbon dioxide, methane, or other carbon gas. The waste product would be passed through a radiation counter. The results appeared quickly. As soon as the first drops of nutrient were injected into the incubator, radioactivity was registered. In 24 hours it rose from a background level of 500 counts per minute to 7500 counts per minute. That count continued to rise at a slower rate to 10,500 counts per minute after seven days, when more nutrient was added.

The response to the additional nutrient did not follow expectations. More of the labeled gas was released for a short time and then the emission decreased. It then began to rise again, but this time slowly. The expectation that the added nutrient would stimulate the exponential growth of a microbial population was not fulfilled. If these were microbes, they were not behaving as Earth microbes would under similar conditions.

Although radioactivity was continuously generated over the 60-day tests, there was no sign of exponential growth. The lack of it created doubt that the test results indicated life, although this was the only experiment that had produced a biological signal that met preflight criteria for a positive response. When the sample was sterilized by heat at the outset, there was no response.

The experiment team considered the possibility that exposure of the soil to solar ultraviolet light had produced the mysterious oxidizing agent hypothesized as responsible for the results of the gas exchange test. The team attempted to check this possibility by maneuvering the scoop to move a rock and dig out soil beneath it. The shielded sample yielded the same ambiguous results as the unshielded sample.

The verdict of the biology team was that the existence of life on Mars, now or ever, was not proved. That conclusion was influenced heavily by the failure of the molecular investigation team's gas chromatograph mass spectrometer to detect organic molecules in the soil. The team tested four soil samples, two at Chryse Planitia and two at Utopia Planitia. One surface and one subsoil sample was taken at each site. The spokesman for the molecular analysis team, Klaus Biemann of the Massachusetts Institute of Technology, reported: "The absence of organic compounds . . . makes it unlikely that living systems that behave in a manner similar to terrestrial biota exist, at least at the two *Viking* lander sites."[20] The mass spectrometer was designed to detect molecules at levels of parts per million to parts per billion. The organic base of the evolution of life was not found in the soil. However, the biology team stated in its summary report: "The amount of carbon fixed in experiment C-1 (the first run of the carbon assimilation experiment at Chryse Planitia) is well below the detection limit of this sensitive instrument."[21] Had the molecular analysis team missed something?

PHOBOS*

In February 1977 the *Viking I* orbiter made two close approaches to Phobos, the nearer and larger of Mars' two moons, and photographed its surface from a distance of 75 miles. Phobos orbits Mars at an average distance of 3750 miles from the surface. Its surface was shown to be cratered and mounded, and the satellite appeared to be an asteroid or a fragment of one. It is shaped like a packing crate, 17.3 miles long, 14.6 miles wide and 12.4 miles high. Its structure has been tentatively determined

*Phobos (*fear*) and Deimos (*flight*) were named by Asaph Hall, U.S. Naval Observatory, who discovered the moons in 1877. They refer to the horses that drew the sun god's chariot across the heavens in Greek mythology. But Homer said they were really the sons or attendants of Mars who hitched up the horses for him.

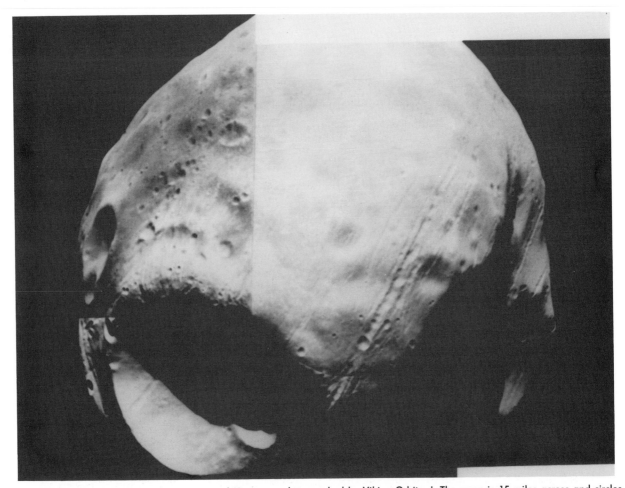

Figure 8.17. Phobos, larger of the two moons of Mars, was photographed by Viking Orbiter I. The moon is 15 miles across and circles Mars every 7.6 hours at 3700 miles altitude. The moon is expected to contain water and has been considered by both the United States and the USSR as a base for exploring Mars. (National Space Science Data Center and NASA)

to be that of a carbonaceous chondrite, a common asteroidal composition that is known to contain a large volume of water. For this reason its potential as an orbital base and refueling station (hydrogen and oxygen propellant) for manned Mars expeditions has attracted the attention of both the United States and the Soviet Union. The smaller moon, Deimos, orbiting Mars at a distance of 12,000 miles has the same composition. It too is box-shaped, with dimensions of 11.7, 7.4, and 6.2 miles.

The *Viking* orbiters and landers were programmed to perform a basic mission from the summer of 1976 to the start of conjunction in November, when Mars moved behind the sun (relative to Earth). After conjunction, observations were extended from year to year, except for the biological experiments which lasted eight months.

Orbiter 2 was shut down July 24, 1978, when its nitrogen attitude control gas was exhausted. *Orbiter 1* continued to photograph the surface and col-

lect radiometric data until August 8, 1980. It was shut down when its control gas ran out.

Lander 2 was switched off in mid-March 1980 because of a power failure. *Lander 1* continued to send photos and weather reports every eight days until it failed March 8, 1983.

AMBIGUITY

Historically, the press briefing by Viking scientists summarizing the results of the search for life on Mars was a memorable occasion. It was held in the ballroom of the National Press Club in Washington November 9, 1976, as Mars entered conjunction, signaling the end of the primary Viking mission. The correspondents learned that the discovery of a new world in the space age is not a event but a process, characterized by ambiguity. The Viking biology team leader, Harold Klein, addressed the

Figure 8.18. Deimos, only about half the size of Phobos, circles Mars every 1.3 days at 15,000 miles altitude. (NASA)

question of ambiguity in the biology experiment result. "I believe that on the basis of the data we have in hand, we cannot conclusively say that there is life on Mars . . . and we cannot conclusively say that there is no life on Mars." Biemann added, "My feeling is that experiments to date have not proved the presence of living systems at the two landing sites. The data (from the biology experiments) are explainable in terms of non-living chemistry. However, I feel that they have by no means excluded the presence of non-terrestrial biochemistry at those places, or even more terrestial-like living creatures at other areas of the planet."

Carl Sagan, Cornell University, leader of the lander imaging team, refused to abandon the rationale for the probability of life. The team leaders—Klein, the biologist; Biemann, the chemist; and Sagan, the astronomer—responded in terms of their disciplines. Sagan reviewed the evidence that supported a belief that Mars had spawned life in a more favorable past environment. The channels and their obvious tributaries showed that liquid water was abundant on the surface a billion years ago, he said. "Now you cannot have liquid water on Mars today because the atmospheric pressure is not high enough and an open pan of water would immediately evaporate and boil away," he said. "To believe in the existence of running water earlier in the history of the planet, a higher pressure atmosphere than now exists is required."

There is some chemical evidence that a denser atmosphere did indeed exist in the past, Sagan said. He cited the relative abundances of argon 36 to argon 38 and to argon 40, which suggest that the early atmosphere had a pressure tens to hundreds of times greater than today—more than enough to make rivers flow. Mars is in an ice age today, he said, but the existence of the channels indicates warmer temperatures a billion years ago. A possible explanation may be found in a theory that Earth's early atmosphere contained 1 part in 100,000 of ammonia. He explained that ammonia is a "greenhouse gas" that would absorb radiation and increase a planet's temperature. Perhaps Mars was warmed in this way.

Going to step further, he explained that it is

known from stellar evolution that our sun is getting brighter with time. Hydrogen escapes and ammonia disappears. The temperature then goes down to the present. Saga applied this process to Mars. If Mars had 1 part in 100,000 of ammonia in a 1-bar atmosphere (the density of Earth's atmosphere at sea level), the average temperature would have been above freezing. "So, putting all this together," he said, "we find that in the early history of Mars, there was all of the atomic constituents necessary for the origin of life—abundant liquid water, higher (atmospheric) pressures and an interesting catalyst."

The catalyst may have been a mineral called montmorillonite, he said. It was found in the Mars soil by the x-ray fluorescence spectrometer team led by Priestley Toulmin III, of the U.S. Geological Survey. Montmorillonite is an abundant iron-rich clay, possibly the most abundant mineral at the *Viking* landing sites, said Sagan. He explained:

The reason that this is of curious relevance to the question of the origin of life is that in experiments performed in the laboratory of Professor Katchalski in Israel it has been found that montmorillonite is a superb catalyst for the combining of amino acids into long chain molecules like proteins. And it turns out it's also catalyst for a number of other synthetic reactions in prebiological organic chemistry cells.

If organic molecules were ever produced in high-yield in early Martian history, the presence of montmorillonite would suggest that they got together to make bigger molecules or even greater elements of the origin of life.

Sagan took issue with Biemann's conclusion that inorganic chemical reactions would explain all the the *Viking* biology results. He asserted that the two biology experiments using radioactive tracers—the carbon assimilation and labeled release experiments—had a higher detection sensitivity than the gas chromatograph mass spectrometer on which the molecular experiment depended to detect organic molecules. Consequently, it was perfectly possible for the biology experiments to detect microbes and for the molecular experiment to miss the organic material in these microbes, he said.

As to the experiments per se, Sagan pointed out

than none was done at Mars ambient temperatures. "It's a little bit like sending a spacecraft from Mars to Earth, but only examining the organisms at temperatures of 200° F." In the same way, he added, the Viking microbiology experiments have been done in the temperature between 50° and 77° F—the highest ever reached on contemporary Mars. "One would dearly like to do such experiments at lower temperatures, maybe even as low as the average Martian temperature to 10 K (−441.40° F).

"Until such experiments are done, it is impossible to exclude the most reasonable sort of Martian life, the kind that likes the ambient temperatures and is destroyed by the high temperatures."

The obvious follow-up to Viking, Sagan said, is an unmanned roving vehicle. It would have improved biological experiments, more reagents, more test scales and onboard logic, a drill to reach depths where ultraviolet light has not penetrated, and a microscope.

A decade after *Viking* fell silent on Mars, data from an elaborate search for life are inconclusive. Extraterrestrial life may be harder to prove than expected, perhaps because biologists are looking for life as they know it.

The results of the *Viking* biology experiments ranged from negative to ambiguous, depending on one's point of view. Conservative biologists viewed the outcome as negative; those less conservative were puzzled by results that failed to prove either the existence or nonexistence of living organisms in the soil samples. It was the Scotch verdict: Not proved. The basic assumption of these experiments held that life would be chemically similar wherever it existed in the Solar System. Processes that would detect it on Earth should detect it on Mars. Consequently, the biology team concluded that life was not found at either Viking landing site. Still, the carbon assimilation experiment had shown that a process analogous to biological digestion and assimilation had occurred. Whether it demonstrated native organic activity or some undetermined inorganic reaction was not determined. Sometime in the 21st century, the United States or the Soviet Union or both will try again and this time, perhaps, drop the other shoe.

The idea that life on Mars would be chemically constructed like life on Earth represented a primary article of faith in biology that biogenic molecules, the seeds of life, have a common origin and are the same among the worlds of the Galaxy of Andromeda as on those of the Commonwealth of Sol. This idea is derived from the belief that all organisms arose by divergence from a common origin, a common ancestry, from which life has evolved into myriad forms, all sharing a common genesis. The chain of events leading to this genesis begins with the Big Bang; it extends to processes operating in the interstellar clouds that spawn stars like our sun, to the transformation in stars of the chemical elements that make up living systems.

This idea promises that life is diffused throughout the universe, that we are not alone in the cosmos. The search for extraterrestrial intelligence draws inspiration from it. It also inspired the *Viking* biology experiments. Was their failure to detect irrefutable evidence of life on Mars, the only other likely abode of life we know of in the Solar System, a result of an inadequate experiment design or a breakdown in the process, or was it a sign that life was not at home when the *Viking* landers called, but might have been in residence on the Martian plains in an earlier epoch.

Although the Viking experience has indicated a possibility that inorganic reactions mimicked organic processes in a way not yet understood, it has suggested that the range of environmental conditions in which life can develop may be more limited than many investigators have supposed. Life may be more restricted and more dependent on specialized environments than the cosmic spawning concept would suggest. On Mars, the biology team theorized, ultraviolet radiation, from which the surface is not shielded, as Earth's surface is by ozone, may have sterilized the planet. Biogenic compounds from which life might have arisen were not found in the *Viking* search. Perhaps they never formed. Perhaps life was not found, unambiguously, because it never got started on Mars. This question is the legacy of *Viking*. It remains for expeditions of the twenty-first century to check it out.

N I N E

THE EMERGENCE
OF MARS

O N May 6, 1983, the Viking 1 lander in Chryse Planitia was pronounced dead by the Jet Propulsion Laboratory. Communication with it was lost the previous November and efforts to reestablish radio contact were abandoned in February. *Lander 2* in the northern plains had failed in 1980; *Orbiter 2* was shut down July 27, 1978, after attitude control gas was exhausted, and *Orbiter 1* was turned off a year later. So ended the most complex and successful planetary reconnaissance of the twentieth century. It left unresolved many questions of planetary evolution, the most tantalizing being whether life had ever evolved there.

Since *Mariner 4* flew by Mars in 1965, leaving the impression of lunarlike desolation, the scientific perception of the red planet matured serially through the fly-by missions of *Mariners 6* and 7 in 1969, the *Mariner 9* orbiter in 1971, and finally *Viking* in the six-year period 1976–1982. From these investigations a planet remarkably similar to Earth in some geophysical aspects and mysteriously different in others emerged.

"Of all the nine planets and their dozens of moons,

Mars still stands out as the one other potentially habitable body in the Solar System," the Solar System Exploration Committee of the NASA Advisory Council stated in its 1983 report.

Venus has been eliminated from consideration by our space exploration. . . . In many ways, Venus is a twin of Earth in size and general composition, but its evolution has produced the very opposite of Earth's surface environment: ovenlike temperatures, crushing atmospheric pressures and a thick, corrosive smog which envelopes the entire plant. These conditions have slowed our investigation of Venus and there is little liklihood that astronauts will ever explore its surface. . . . From our present perspective, the only planet in the Solar System that we can hope one day to understand as well as our own and perhaps live on as well as our own is Mars.

By the mid-1980s Mars became established among space planners as the primary focus of space exploration, as the Moon had been in the 1960s. As NASA sought to recover momentum after the *Challenger* accident of 1986, Mars emerged as the postlunar goal of a new and bold initiative. For the

first time since Apollo, planning manned expeditions beyond low Earth orbit won the support of the NASA bureaucracy. That had not been evident before the accident, when the agency's focus was limited to low Earth orbit and the development of the reusable space transportation system. Although this outlook encompassed the space station, the role of a facility dedicated to Earth observation and microgravity experimentation was not quite enough to capture the imagination of the space-interest community. While the shuttle was grounded during the redesign of its solid rocket boosters, the space station concept was enlarged to project its use as a base station for manned and robotic exploration of the Solar System.

By 1987 manned expeditions to Mars, which had been analyzed for years in semiobscurity at NASA centers, were added to the agency's agenda and confirmed by a brief statement in President Reagan's new space policy calling for a manned presence beyond the Earth. The presidential dictum appeared to have overridden the influence of scientists principally in academia who opposed costly

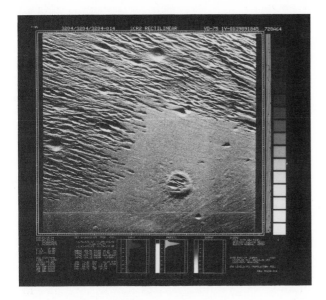

Figure 9.1A. Sand dunes on Mars were photographed by Viking Orbiter I. These dunes have been elongated by the winds into spectacular formations. Origin of the sand grains has puzzled investigators inasmuch as wind blown dust is considered too fine grained to have formed the dunes. Dune formation, it is believed, requires coarser material. One theory holds that dust particles are clumped together electrostatically or bound together by ice into sand-sized grains. (NASA-Viking).

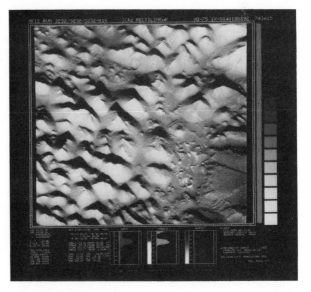

Figure 9.1B. Chaotic ground on Mars, the product of wind and possibly ancient water ice erosion. Frost has formed on the hills, leaving the ravines in shadow.

manned ventures as less productive scientifically than unmanned probes but more likely to get the lion's share of NASA funding. It was an old and irreconcilable conflict.

A new Department of Exploration to coordinate manned expedition planning was installed at NASA headquarters. The shift in focus was the bureaucratic reponse to the despair generated by the fate of *Challenger*. Moreover, it had widespread public support and was backed by the National Commission on Space.

In 20 years of reconnaissance, the image of Mars

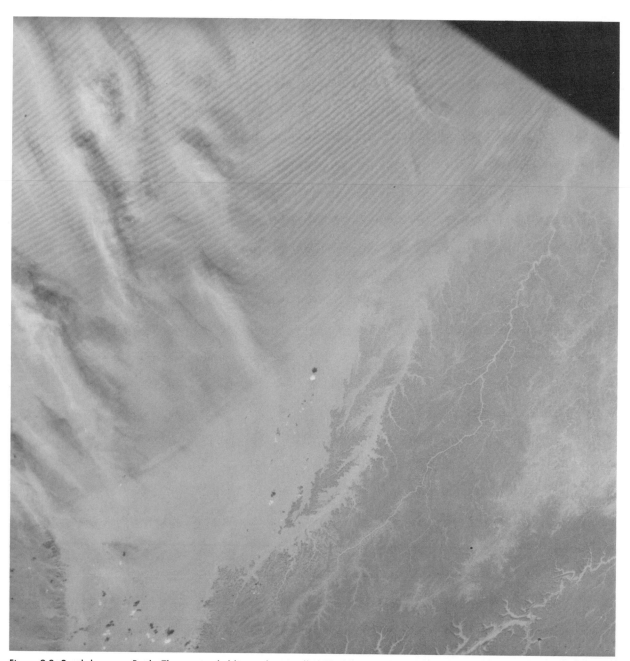

Figure 9.2. Sand dunes on Earth. These extended linear dunes called "Seifs" were photographed from low Earth orbit on the first flight of the shuttle Columbia Apr. 12–14, 1981. They extend for hundreds of miles in Saudi Arabia and Yemen, in a region called the "Empty Quarter." (NASA)

had been transformed from that of a featureless desert to a land of scenic grandeur. The Solar System Exploration Committee's 1983 report described individual shield volcanoes that would stretch from Boston to Washington, a grand canyon extending as far as the distance from Los Angeles to New York and a sea of sand dunes girdling the Martian arctic. Moreover, the report said, "there is ample evidence of massive climate changes on both planets—ice ages, changing sea shore lines and species extinctions on Earth; regional flooding, glaciation and polar sedimentary layering on Mars. Common mechanisms could have been at work on both planets: solar luminosity changes, periodic orbital variations, episodes of volcanic eruption and asteroidal impact."

All these features were clues to the fundamental processes of planetary formation and evolution that remained to be investigated further now that radio silence had returned to Mars with the demise of *Viking*. The Russians sought to break it with Phobos reconnaissance probes launched in 1988, but their radios failed. Still, the legacy of *Viking* had left science with the question of whether there was or ever had been life on Mars. Could robots resolve that issue?

TO BE OR NOT . . .

The *Viking* biology experiments left a mystery that can be pondered but not satisfied by the results. At least one of the three life experiments conducted robotically in the minilaboratories of the two landers produced results that seemed to be evidence of microbial life. It was the labeled release experiment. Of the other two, one was inconclusive and the other negative. The question of life on Mars remained open. Central to it was the failure of the Molecular Analysis Team to detect organic compounds in the soil. Their absence cast doubt on the existence of living systems, at least at the *Viking* lander sites, said the team's spokesman, Klaus Biemann of the Massachusetts Institute of Technology.

However, Biemann added a qualification to the evidence on which he based the likelihood that radiation destroyed organic compounds on Mars as fast as they formed. "The *Viking* gas chromatograph mass spectrometer was not designed to search for life on Mars," he said. "Consequently, the demonstration that very little, if any, organic material is present does not exclude the existence of living organisms in the samples analyzed and certainly

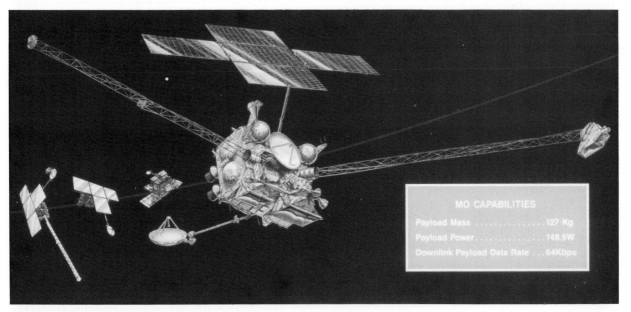

Figure 9.3. The basic initial Mars reconnaissance spacecraft is the Mars Observer, carrying 280 pounds of instruments. The Observer is similar to the polar orbiting TIROS satellites with years of meteorological service in U.S. meteorological programs. The payload includes a high resolution imaging system, gamma ray spectrometer and magnetometer. (Pathfinder)

Figure 9.4. A step up in design from the Lunar Rover in Project Apollo, this artist's conception of a Mars Rover is designed for automated mobility on the surface of Mars as well as for control by a human driver. It would collect surface samples robotically. (Pathfinder).

does not rule out the possibility of a rich biota out of range of *Viking* landing sites. Material taken from very different sites, for example, the polar regions where organic substances have been cold trapped would be of much greater interest."[1]

The door was open to go back to Mars and look again. The Solar System Exploration Committee expressed a cautious view. "The Viking results," it reported, "make it hard to sustain the notion that present day life exists on Mars, even though the basic ingredients—water, carbon compounds and energy—are all there. Life may have developed on Mars in the distant past when liquid water may have periodically flooded the surface and when precipitation almost certainly took place, but this separate issue will not be settled until samples of Martian rocks can be studied in detail. The recognition of fossils in the rocks such as the layered sedimentary units seen on both the equatorial and polar regions of Mars would be a discovery for the ages."

THE WAY BACK

A resolution of the life mystery in the reconnaissance phase of Mars exploration would require a mission that would land on Mars and return samples of the soil and rock at several locations to Earth. Initially proposed by the Solar System Exploration Committee in 1983, the project was combined with a roving vehicle to provide mobility for the collection of samples and evolved as the Mars Rover/Sample Return Mission.

As defined by the Solar System Exploration Committee, the objective of the sample return mission was to collect documented samples from a known location.[2] This was necessary to understand the origin of the planet and its chronology. Sample analyses would be expected to shed light on two questions: the past or present existence of life and a possible Mars origin of a group of rare meteorites with properties suggesting a planetary origin.

In the years following the Viking project, the importance of a Mars sampling expedition became increasingly obvious as a means of determining the evolution of the planet and and its climatic history. Studies of a sample return mission were done at the Jet Propulsion Laboratory, at the Johnson Space Center, and by a consultant, Science Applications International. These concluded that a 5-kilogram (11.2 pounds) sample could be collected by an automated sampling device mounted on a roving vehicle, lifted to Mars orbit, and sent back to Earth. Sufficient data from *Viking* were available for the selection of a landing site. The rover would have a range of several tens of miles. The entire process could be automated.

While an 11-pound return from Mars would be hardly comparable to the samples brought back by Apollo expeditions (a total of 833.9 pounds), it would be considerably bigger than automated moon soil returns from the Soviet *Lunas 16* and *20,* which brought back a total of about 7 ounces in 1970 and 1972. Defeated in the race to land the first men on the Moon, the Soviets created an automatic sampling system that has become the precursor of a rover/sample return mission now being considered by both space-faring powers for Mars.

With the Mars sample, the Solar System Exploration Committee hoped, an absolute chronology for the rocks could be established by the analysis of long-lived radioactive elements and their daughter products. These would provide isotopic ages (the decay time) of the rocks in the sample. The committee suggested that samples be collected from volcanic regions, layered units, and polar regions. These would include fresh igneous rocks, weathered igneous rocks, breccias (consolidated rock fragments), loose debris, and possibly some trapped atmosphere. The space station is not necessary for the sample return mission but could provide facilities for recovering the sample and for preliminary examination in Earth orbit. The Mars *Observer* satellite might identify sample sites, although its main function has been designed to survey the mineralogical character of the surface material, the distribution of elements, the topography, and the magnetic and gravitational fields. Development of the *Observer* had been started in 1984 and its launch was planned in 1990. In addition to chemical analysis of the soil, it would map the planet for a year and transmit data on the structure and circulation of the atmosphere.

From the *Mariner 9* orbiter and *Viking* orbiters,

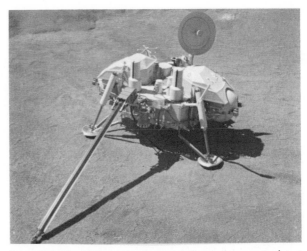

Figure 9.5. The Viking I Lander, 1976. Its mission was to detect evidence of life on Mars. The ambiguous results have stirred years of controversy. (NASA).

NASA had acquired global photomaps with a resolution of less than a kilometer. Those missions had produced localized images covering some areas less than that of a football field. The *Viking* orbiters had radioed photos with a resolution of 50 to 75 feet in anticipation of selecting a landing site for a sample return. The Mars *Observer* mission would then support the final choice for a landing site, the committee said.

The committee considered the rover as a mobile platform for the tools that would obtain the sample. These were a drill, a core tube, a sampling arm that would pick up loose fragments and sand, and a camera to observe the collection process.

Figure 9.6. The Ladokh and Zaskar ranges of the Himalayas, etched by snow and shadow on the border of India and China, as photographed by the Space Shuttle Columbia, 1981.

In this contest the rover was considered only as a sampling device. No extraneous experiments, using meteorlogical or astronomical instruments, should be added to it. In addition to the mechanical collection apparatus, it would be equipped with stereo cameras capable of recording and transmitting multispectral images, with a device to make simple chemical tests, a Geiger counter to detect radioactive materials, and a lifting device to measure soil weight and density.

If the sample was to be collected at a single place, such as one of the *Viking* lander sites, the rover should be able to travel 300 feet. The committee reviewed the experience of the astronaut-operated *Apollo 15* rover, the first American wheeled vehicle on the Moon, in the summer of 1971. The lunar jeep traversed a wide area at the foot of the lunar Apennine Mountains and along the Hadley Rille. The automated rover on Mars would move more slowly, directed from Earth. Most of the mission time would be spent in the sampling operation. The committee study calculated that actual travel time for a traverse of 155 days on Mars would be only 31 hours at an average speed of 10 centimeters (25.4 inches) a second. This was based on a simulated mission.

If the sample was to be collected from only one site, the committee estimated that it would take several months to pick up 8.8 pounds, with the rover ranging over a 300-foot radius. Time would be spent in planning the moves on Earth as well as in the actual sampling operations.

A baseline sample mission was devised in a 1984 study by the Jet Propulsion Laboratory, the Johnson Space Center, and Science Applications International with the launch set for 1996. It called for placing a Mars Vehicle system with a mass of 20,000 pounds in low Earth orbit by the shuttle. The Vehicle would be mated there with a Centaur rocket orbited by a second shuttle flight. The Centaur would boost it on a 303-day flight to Mars orbit and separate.

As the Mars Vehicle encountered Mars, it would automatically execute an aerocapture maneuver by passing through the upper atmosphere long enough to reduce its interplanetary velocity and allow it to fall into an elliptical orbit, 348 by 1242 miles, around Mars. A Mars Orbiter and Earth Return Vehicle would then separate from the vehicle system and remain in the parking orbit while a lander in a Mars Entry Capsule descended to the surface with the roving Mars Rendezvous Vehicle vehicle and sample collection device. The Entry Capsule/lander would apply aeromaneuvering techniques to fly the cargo to the landing site. On final approach, a parachute would be deployed and a terminal descent engine would ignite to make a soft landing.

Approximately 11 pounds of soil would be collected by the rover–sampler and sealed in a canister emplaced in the Mars Rendezvous Vehicle. This element would be lifted to the parking orbit after a stay time on Mars of 401 days. The Rendezvous Vehicle would dock with the Mars orbiter and the sample would be transferred to the Earth Return Vehicle. This element would be separated from the Mars Orbiter and launched on a 326-day trip back to Earth. Total mission time would be 1030 days.

This scenario does not require the use of a space station, but if one is operational at the time of the mission, it would be useful. With the station in orbit, the Centaur could be orbited dry and loaded with hydrogen/oxygen propellant at a station complex. Alternatively, an Orbital Transfer Vehicle garaged at the station might be used to move the Mars Vehicle to Mars orbit.

The sample collection mission scenario implies advanced automation and robotics as well as detailed radio control and monitoring from Earth. The return calls for an aerocapture maneuver to reduce interplanetary velocity, to let the Vehicle and capsule enter a 12 hour orbit. Its perigee reaches the orbit of the space station or the shuttle so that the capsule can be recovered. To eliminate the risk of contamination, preliminary sample analysis would be done in a space station laboratory.

This sample return scenario assumed the use of the Centaur, the most powerful upper stage in the American inventory for the Earth orbit to Mars orbit transfer. After the *Challenger* accident, NASA banned the transport of the Centaur in the shuttle cargo bay because of the explosion potential of its hydrogen–oxygen propellant. If Centaur was to serve as an interplanetary booster, it would be lifted to low Earth orbit dry or by a Titan 4. At the begin-

Figure 9.7. Four types of Earth to Mars transfers have been charted by Science Applications International Corp., Schaumburg, Illinois (SAIC):

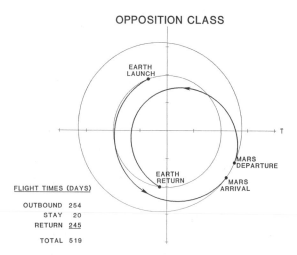

OPPOSITION CLASS

FLIGHT TIMES (DAYS)

OUTBOUND 254
STAY 20
RETURN 245

TOTAL 519

Figure 9.7A. Opposition class missions when Earth and Mars are on the same side of the sun at launch allow a 20 day stay at Mars and a relatively quick passage there and back to Earth. (SAIC)

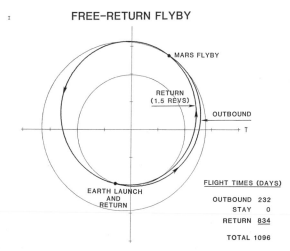

FREE-RETURN FLYBY

FLIGHT TIMES (DAYS)

OUTBOUND 232
STAY 0
RETURN 834

TOTAL 1096

Figure 9.7B. A free return fly-by allows a short passage to Mars but a long one home. (SAIC)

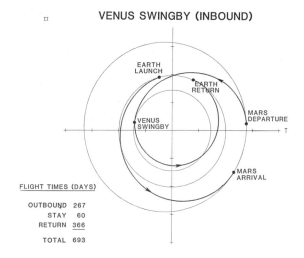

VENUS SWINGBY (INBOUND)

FLIGHT TIMES (DAYS)

OUTBOUND 267
STAY 60
RETURN 366

TOTAL 693

Figure 9.7C. A homebound swing-by of Venus allows 60 days at Mars. The homebound passage is accelerated by gravity assist from Venus. (SAIC)

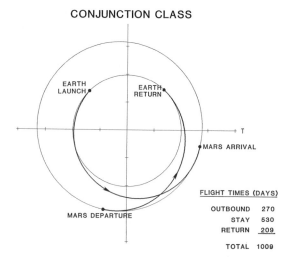

CONJUNCTION CLASS

FLIGHT TIMES (DAYS)

OUTBOUND 270
STAY 530
RETURN 209

TOTAL 1009

Figure 9.7D. Conjunction class missions when Earth and Mars are not on the same side of the sun allow a 530 day stay at Mars and 7 to 9 months passage. (SAIC)

ning of 1988, Air Force Titan 4 launchers began arriving at Cape Canaveral. They were billed as capable of lifting as much tonnage to low Earth orbit as the shuttle and hence were described as shuttle-class expendable rockets.

A PIECE OF MARS

In 1987 a revised Mars sample return initiative was described by Science Applications International scientists at the AAS/AIAA *Astrodynamics Specialists Conference at Kalispell, Montana, August 10–13. It was titled the Mars/Rover Sample Return Mission to denote its two separate mission elements: the rover and the sample return system. The initiative grew out of the 1984 study with the participation of engineers and scientists at the Jet Propulsion Laboratory, the Johnson Space Center, the Ames Research Center, the U.S. Geological Survey, and Science Applications. NASA headquarters created a Mars Exploration Science Advisory Group to provide oversight and guidance for the project.

The reference launch date for the mission was 1998. By splitting it into two discrete elements, the planners provided for international participation "with minimum technology transfer and maximum sharing of results," the Science Applications paper said. The scenario here called for a minimum energy conjunction class flight to Mars of 289 days. Both the rover and the sample return lander vehicle system would be launched at the same opportunity "window" to reach Mars within a month of each other. At encounter, each system would be inserted into Mars orbit. Mission time at Mars would be 489 days and the return of the sample to Earth would take 220 days. Total mission time would be 998 days, or 2.7 years. With a launch at the 1996 opportunity, arrival would occur during the fall season in the northern hemisphere, when atmospheric conditions would be expected to be calm. However, a reconnaissance for Mars' violent dust storms would be carried out before the rover and

*American Astronomical Society/American Institute of Aeronautics and Astronautics

the sample return vehicle would be committed for landing. This could be done by a Mars *Observer* satellite or by imaging devices provided for the mission while the interplanetary transfer vehicle is orbiting the planet.

An alternative to direct launch from low Earth orbit was considered—a swing by Venus. While flying to Mars by way of Venus would appear to be a detour, it would reduce total trip time for the mission. The swing-by of Venus would increase spacecraft velocity with the gravitational assist of that planet, but the maneuver would require higher propulsion to achieve escape velocity on the Earth-to-Venus leg of the trip. By applying the Venus swing-by, controllers could reduce total mission time from 992 to 729 days, cutting the stay time at Mars from 332 to 60 days. If the conjunction class mission departed Earth November 17, 1996, it would not return to Earth until August 6, 1999. On the Venus swing-by, a mission departing March 16, 1996 would be back March 16, 1998.

The components of the rover/sample return systems could be carried to Mars together on a single launch from Earth orbit or as two payloads on separate flights. An integrated system would have an orbiter, Earth Return Vehicle, Earth Orbital Capsule, Mars Ascent Stage, the Rover, the Mars lander, and an aeroshell to shield the vehicles during the aerocapture maneuver. All seven elements would have a total mass of 8000 kilograms (17,640 pounds). It would require a Saturn 5 class heavy-launch vehicle to launch them directly to Mars from Earth.

The dual launch splits the Mars payload into two packages: sample return and rover. It could use available launchers—the shuttle with an intertial upper stage (IUS) and the Titan 4 and Centaur upper stage. The packages would be launched separately during the same Mars launch window. The landers would be deployed from their respective orbiters in Mars orbit and the sample return machines would make their rendezvous with the rover. All operations would be monitored from Earth through orbiter communications relay.

The package containing the 700-kilogram (1543-pound) rover would be launched from Cape Canaveral by a Titan 4 Centaur and proceed to Mars carrying the orbiter, lander, and protective aero-shell. The total mass would be 2200 kilograms (4850 pounds). The larger of the two packages, the sample return machine, would have an Earth launch mass of 5800 kilograms (12,786 pounds) with its orbiter, Earth Return Vehicle, Earth Orbital Capsule, Mars Ascent Stage, lander, and aeroshell. It too could be sent to Mars by Titan 4 Centaur or, alternatively, by the shuttle–IUS to low Earth orbit and from there to Mars by a Centaur placed in orbit by Titan 4. A staging operation would be required, probably involving shuttle astronauts, to transfer the sample return payload to the Centaur.

The scenario developed for the dual payload launches calls for the separate payloads to be injected into an orbit around Mars by aerocapture or retropropulsion at encounter, as follows:

The rover enters an elliptical orbit with a periapsis altitude of 155–310 miles and an orbital period of one Martian day (24 hours, 37 minutes). The sample return payload may be placed in a similar orbit. The two landers are then de-orbited to enter the atmosphere. They are guided to landings near each other, the sample return system landing first, followed soon by the rover. Sampling is done primarily by the Rover, although provision may be made for some sampling by the sample return system.

Stay time on the surface was calculated at 11 to 18 months. When sampling is completed, the rover returns to the sample return system lander and transfers its loaded canister to the Mars Ascent Vehicle. The Ascent Vehicle lifts off, using the lander as its launch platform, and makes rendezvous with the orbiter which carries a separable Earth Return Vehicle. The sample canister is transferred to the return vehicle which fires its rockets to return to Earth. As it approaches Earth, the vehicle ejects a capsule containing the canister into an orbit, where it is recovered by the shuttle.

What comes back from Mars is a 112 pound capsule containing a 41 pound canister with 11 pounds of Mars dirt and rocks in it. The mission cost was estimated at $3.5 to $7 billion.

LUNA 16 AND 17

The Russians were first to return a soil sample automatically from another body—the Moon. They put the first automated roving vehicle on the Moon, although it was not part of a sample return effort. That came a bit later. The motivation for these technical feats was the Soviet inability to put men on the Moon with bucket and shovel. They resorted to automation, with conspicuous success.

While these efforts were regarded in the early Apollo period as a political response to the American manned landings on the Moon, they were also recognized in the American space community as prototype systems for planetary exploration. The Soviet sampling and rover missions to the Moon were designated as *Luna 16* and *17*. They were successful and later repeated. *Luna 16*, the sampler, was launched September 12, 1970, into a parking orbit from which the sampler vehicle was boosted Moonward, leaving behind its carrier cylinder and orbital platform. The carrier body was initially estimated by British observers as 39 feet long and 13 feet in diameter with a mass of 8820 pounds.

Following a midcourse correction September 13, the sampler fired braking rockets September 17 and dropped into a circular orbit of the Moon at 68 miles altitude. The retrorockets were again fired to break the sampler out of orbit and control its descent to the surface. The main rocket engines were turned off at 65 feet and two vernier engines at 7 feet. Luna 16 landed with a bump in the Sea of Fertility at 0 degrees, 41 minutes south latitude and 56 degrees, 18 minutes east longitude.

The sampler vehicle consisted of a stack of spherical tanks and cylinders held together by open truss work. Four shock-absorber legs extended from the base and the main descent engine nozzle protruded from the center, where the liquid fuel descent engine was located. Atop the four-legged landing stage sat the ascent stage, with its own liquid fuel main engine and vernier engines. The recovery sphere designed to contain the sample canister crowned the sampler like a dome. It was protected from the heat of reentry at Earth by ablative shielding. In addition to the sample container it enclosed parachutes, a radio beacon, batteries, and antennas.

As landed on the Moon, the sampler had a mass of 4145 pounds, was 13 feet high and 13 feet across. An arm attached to the landing stage was projected beyond the blast area of the descent rockets and held a drill consisting of a hollow tube with cutters at the end. This rig was designed to extract about 100 grams of rock and soil at a depth of about a foot. Sensors on the drill enabled controllers on Earth to regulate the cutting speed and determine the hardness of the ground. When the sample was collected, the sampler arm carried the core to the ascent stage and inserted it into the recovery capsule, where it was sealed.

After 26 hours and 25 minutes on the surface, the ascent stage bearing the return vehicle with the capsule lifted off the lander and flew directly back to Earth. The recovery capsule was separated from the Earth return vehicle by an explosive charge and plunged into the atmosphere. Heavily insulated, it survived frictional heating up to 10,000 degrees C. A braking parachute was deployed followed by the main parachute and the radio beacon antenna. Guided by the beacon, pilots in fixed-wing and helicopter aircraft sighted the descending capsule, and helicopter crews recovered it.

The total weight of the Moon sample was 101 grams, about 3 ounces. It was flown in the sealed container to the Lunar Receiving Laboratory in Moscow and placed in a stainless steel chamber with glass portholes. After the chamber was sterilized by gas, the container was opened by remote control. Portions of the small sample were passed to sublaboratories equipped with manipulatoers and built-in rubber gloves that allowed investigators to work directly with the sample. Similar facilities were installed in the NASA Lunar Receiving Laboratory at the Johnson Space Center, Houston, where 833.9 pounds of lunar rocks and soil were examined in a quarantine laboratory environment. Returning lunar crews were quarantined in the laboratory on their return until 21 days had passed since their departure from the Moon. The 21-day crew quarantine was discontinued after it was completed by

Alan B. Shepard, Jr., Stuart A. Roosa, and Edgar Mitchell on their return from the Apollo 14 mission to the Fra Mauro Formation in February 1971. After three crews and bags of lunar samples had returned from the Moon without any sign of an organism, living or fossil, Apollo life scientists concluded that life had never existed on the Moon before men reached it.

The Soviet rover mission, *Luna 17,* was launched November 10, 1970 and landed in the Sea of Rains (38 degrees 17 minutes north and 35 degrees west) November 17. The landing stage carried a flat platform with ramps fore and aft on which the rover, *Lunokhod* I, was mounted. There was no ascent stage. *Lunokhod* was meant to stay on the Moon forever after performing its mobile reconnaissance.

The vehicle was shaped like an old-fashioned bathtub with eight wheels, four on a side. The tub section was covered by a convex lid which could be opened out to expose photovoltaic cells to convert sunshine into electricity during the lunar day and to recharge batteries. During the lunar night the lid was closed and the tub was heated by a radioisotope heater. The peripatetic bathtub had a mass of 1667 pounds.

Lunokhod was controlled from Earth by a four-man crew. The electric motor drive had two speeds in forward and reverse. The tub carried antennas, television cameras, an x-ray spectrometer to analyze soil constituents, a cosmic ray detector, and an x-ray telescope. It mounted a device to exert impact force on the soil as a means of measuring its density and other mechanical properties. Automatic sensors would halt the drive when they detected deep holes or steep slopes in the vehicle's path.

Lunokhod also carried a laser mirror. A similar reflector was set up at lunar landing sites by Apollo astronauts. By reflecting laser beams aimed from telescopes on Earth, the mirrors made it possible to refine the distance between the Earth and the Moon and measure the motion of the crustal plate on which the terrestrial observatory was located.

By January 18, 1971, *Lunokhod* had traveled 11,677 feet. Active research operations ended October 4, 1971. A second rover, *Lunokhod 2 (Luna 21),* landed January 16, 1973, in the LeMonnier crater on the rim of the Sea of Serenity and performed a successful reconnaissance of 23 miles. Two more sample return missions, *Luna 20* in 1972 and *Luna 24* in 1976, returned 1.62 and 5.47 ounces of lunar samples to the U.S.S.R.

THE MARS INITIATIVE

It took a decade for the latent initiative to explore Mars with robotic and manned expeditions to be adopted in NASA planning. The idea of going to Mars acquired substance year by year at NASA centers, among scientists at other government laboratories, and at academic and consulting research agencies that were developing to support NASA. The private agencies, acting as consultants, appeared as a response to the stasis in planning and stultification of initiatives that had characterized NASA headquarters after the retreat from the Moon in 1972 and the subsequent abandonment of *Skylab* and the Apollo Saturn transportation system.

A Mars initiative became feasible with the development of the space shuttle, and the presidential decision to build a permanent space station in low Earth orbit, and studies that showed it was necessary and well within the state of the art to build an Orbital Transfer Vehicle. This vehicle, essentially an extension of upper-stage rocket technology, would operate only in space, from the space station in low Earth orbit to lunar orbit or to geostationary orbit. The OTV, as the planners referred to this upper stage, and a revival of the Apollo-style lunar lander would make it possible to return to the Moon and set up a base there. Once the infrastructure of an Earth-Moon transportation system was developed, it would form the matrix for an interplanetary transportation system. Propulsion could be provided by a cluster of OTVs. The space ship going to Mars would consist of space station habitat, laboratory, and logistics modules, with a lander that would ferry crew and supplies between the space ship in Mars orbit and the surface.

Thus, with a shuttle-serviced space station operating in low Earth orbit by the middle or late 1990s, NASA would have an orbital base from which to return to the Moon and explore the Solar Sys-

tem. The vehicles that would effect the transfer to the Moon or Mars remained to be built, but the technology was straightforward and well known. By 1984 the entire infrastructure of the interplanetary transportation system that would allow humans to establish a presence in the Solar System had been meticulously detailed by hundreds of American engineers and scientists.

It remained to be adopted and implemented by government. This was done in stages, each marked by concerns about cost. Responding to criticism of NASA's failure to set long-range goals in space, Congress created the National Commission on Space. Its mission was essentially to drum up public support for an ambitious manned space program and define it in detail. As mentioned earlier, this was done in a series of public hearings starting in 1985. Proposals reflected the mood and content of those drafted by NASA in 1969 for President Nixon's Space Task Group. Mars remained a major objective.

Made public in the spring of 1986 a few months after the *Challenger* disaster, the report of the National Commission was the first document of national scope to assert that the heritage of mankind was the Solar System, with all of its vast resources of matter and energy, not merely the Earth, with its constraint of limits to growth. The report projected a bright new vista for a human future in space, at a time when the public perception of NASA's capabilities and prospects had reached its nadir. The goal of 21st century America in space would be no less than the development of the solar frontier to make accessible new resources of matter and energy and to support human settlements "beyond Earth orbit from the highlands of the Moon to the plains of Mars," said an epigraph.*

So would commence, in the commission's view, the expansion of the human species into the Solar System. The expansionist themes in the commission's report were set forth in bold detail in the report by Astronaut Sally Ride and her group at NASA headquarters in 1987. The Ride report had been commissioned by the NASA administrator,

---

*The phrase quoted exhibits poetic license inasmuch as the Moon is not beyond Earth orbit but in it, accompanying Earth around the sun.

James C. Fletcher, a Republican. Its proposals matched those of the National Commission and its chairman, Tom Paine, a Democrat. It appeared that at this juncture, the programs outlined by both documents for the 21st century had bipartisan inspiration.

Meanwhile, groups of engineers and scientists at NASA and other research centers had formulated the way missions to the Moon and Mars could be done. The exploration of Mars had been analyzed in considerable detail and seemed entirely feasible.

A GRASS ROOTS PROGRAM

Following the first national Conference on Lunar Bases and Space Activities of the 21st Century in 1984, a study of manned missions to Mars was undertaken by a working group whose number swelled to 140 scientists and engineers by 1985. One of the prime movers was former U.S. Senator Harrison H. (Jack) Schmitt (R–N.M.), who as an astronaut explored the Taurus–Littrow region of the Moon with Eugene A. Cernan and Ronald E. Evans on the Apollo 17 mission in 1972.

The members of the working group represented 26 government, academic, and private research agencies ranging in scope from the Goddard Space Flight Center to the Central Intelligence Agency. It was assembled through various technical and scientific meetings beyond the pale of NASA headquarters direction or policy.* The group held a series of meetings in 1985 at the Los Alamos National Laboratory, the Johnson Space Center, the Kennedy Space Center, and the Marshall Space Flight Center. From these session evolved the most comprehensive analysis of manned interplanetary flight of the space age.

The Working Group report, edited by Michael B. Duke of the Johnson Space Center and Paul W. Keaton of the Los Alamos National Laboratory, stated in its preamble, "It is believed that most Americans

---

*In 1983 the Solar System Exploration Committee of the NASA Advisory Council had called for a "core program" of planetary exploration of high scientific priority but "moderate technological challenge and modest cost." The core program did not encompass manned interplanetary missions.

would find it unacceptable for political, economic or intellectual reasons for some other nation to dominate this effort. ... It must also be emphasized that the potential for international cooperation on a Mars venture is real and has far-reaching implications." This sentiment, like the report itself, came from the grass roots of the space technology community in America. It was echoed in the National Commission and Sally Ride reports.

Assessing the space flight state of the art in America, the report concluded that manned flights to Mars were well within the capability of present propulsion and life support engineering. A preliminary estimate of the cost of an initial Mars expedition forecast a bill of $30 to $40 billion. This amounts to about one half of the cost of Apollo, according to the working group's summary report (1986). The group asserted that the lunar missions were developed and flown when the gross national product was one third as large as it was in 1985.

As in Project Apollo, preparation for a manned expedition to Mars required reconnaissance. This had been done by *Mariner* and *Viking,* but more was needed, especially samples. Mars studies, the report said, include chemical comparisons of its rocks with those of Earth and the Moon. Like Earth, Mars has sustained volcanism, but not the tectonic activity that has processed and reprocessed Earth's crust.

Although the *Viking* mission failed to prove the existence of life on Mars, it amplified the possibility that life might have started there when climatic conditions were more favorable in the far past. From a biological as well as a philosophical viewpoint, the existence of life, past or present, is the main focus of exploring Mars. Understanding the evolution of the present environment of Mars has a bearing on the prospect of habitation there. Mars was wetter and its atmosphere denser in the past, the photo evidence has shown. It may well have been more hospitable to life, the group said. "These limited indications that life could have existed on Mars suggest that the habitation of Mars may be feasible," the report said. "Eventually, self-sustaining colonies may evolve there."

The working group said it appeared unlikely that any major scientific issue will be resolved in Mars reconnaissance prior to the first human exploration of the planet. While the Mars Observer satellite and the sample return mission would provide information on the chemical composition of the Martian regolith, human expeditions would be required to allow extended investigation of the nature, evolution, and dynamic processes on Mars. These investigations would develop the means for sustaining and protecting humans on Mars, locate sources of water, and lead to geological and engineering studies to support surface operations, the group said. Provision would be made to monitor solar flares and assess the radiation environment. The early expeditions would test the performance of the closed ecological life support systems that would be a basic requirement for sustaining people on Mars.

THE SCALE OF MISSIONS

The working group considered a range of missions. For a Mars fly-by and moons reconnaissance, the spacecraft and instrument packages would be small and self-contained. For missions that land people on the surface, the major requirements are habitation and surface transportation. For short stays, the lander would provide sufficient shelter. A short-range vehicle like the lunar jeep used in Apollo landings would provide sufficient transportation, mainly short-range reconnaissance to select a site for a long-term base.

A permanent outpost that would sustain people for years would require more complex power and life support systems. It would depend to some extent on local resources, particularly water. The group envisioned large, pressurized vehicles to explore areas of tens to hundreds of miles.

At the outset of establishing the permanent base, the use of indigenous resources for support of the base would be tested. The initial habitat will be carried from Earth, but as the base develops shelters could be constructed of Mars material. The life support system should make use of the atmosphere for the extraction of oxygen (from carbon dioxide) and test the soil for agriculture (under a pressurized glass enclosure). Propellant manufactured from the atmosphere or surface materials will reduce or

eliminate the need of bringing propellant from Earth. Tunneling may be used to protect the long-term habitat and work space that must be shielded from radiation from the sun and other stars.

In many respects, a base on Mars would be developed in much the same way as a base on the Moon. Both would require pressurized, radiation-proof shelters that could be constructed of local material. Both would require ground transport vehicles running on electrical power. On Mars, however, a light airplane could be designed to fly in the thin atmosphere, the group suggested.

Conjunction and Opposition Missions. Mars flights would be launched from low Earth orbit. The launch site would be a specialized base for interplanetary missions or a generalized space station. Propellant would be stored and vehicles assembled at the orbital base and crews and supplies would be ferried up to it by the shuttle.

Two classes of missions were studied by the working group, each characterized by the relationship of Earth and Mars at particular times. Shorter missions are possible when Mars is at opposition—that is, on the same side of the sun as the Earth. When not on the same side of the sun, the planets are said to be in conjunction. A conjunction mission takes longer and allows a longer stay on Mars. It can be flown more cheaply—at minimum energy.

An opposition mission using liquid hydrogen–liquid oxygen propellant has a round-trip time of two years with a 60-day stay on Mars. With this (cryogenic) propellant, the conjunction class mission has a total trip time of three years, with a little more than one year on Mars.*

A Mars fly-by mission allowing just a look-see can be done in one year. A working group scenario charted a Mars fly-by at opposition in 1999. The ship is launched from Earth orbit April 2, 1999, passes Mars August 8 that year, and arrives at Earth orbit April 2, 2000. The working group said that the fly-by mission is "spartan and yields only a few minutes of high speed passage near the planet

---

*Manned Mars Missions working group estimates.

Mars, with no opportunity for a surface visit by the crew members."

An opposition mission of less than two years can be accelerated by the gravity of Venus if the space ship swings by Venus on the outbound leg. The working group showed that if the space ship departed Earth January 26, 1998, and passed Venus July 9 that year, it would arrive at Mars in less than a year, January 16, 1999. With a stay of two months, it would depart Mars March 17, 1999, and reach Earth orbit eight months later, November 16, 1999. Total trip time would be less than 23 months.

A conjunction class mission departing Earth December 17, 1998, would arrive at Mars orbit September 28, 1999, after a voyage of a little more than 10 months. The crew would remain at Mars 16 months until departure January 25, 2001, and return to Earth September 2, 2001, after a nine-month trip home. Total trip time is less than three years. In terms of time away from home, the Mars conjunction flights are analogous to antarctic expeditions in the first two decades of the 20th century.

A conjunction flight allows the space ship to make a Hohmann transfer, an elliptical path from one planet to another requiring minimum expenditure of energy. The Hohmann Transfer Ellipse was described in 1925 by Walter Hohmann, city architect of Essen-on-the-Ruhr, and published in a book entitled *Attainability of Celestial Bodies*. The transfer works ideally for two planets traveling in circular orbits in the same plane.[3]

The spacecraft leaves one planet in a direction tangential to its orbit and encounters the other planet tangentially to its orbit. The flight path is one half of an ellipse around the sun. Earth and Mars generally fit this prescription, but not ideally. Neither moves in a circular orbit—the orbit of Mars is highly elliptical—and their orbits are not precisely in the same plane. For an idealized Hohmann trajectory, the transit time from Earth to Mars is calculated at 260 days. This route becomes possible, with certain allowances, when the relative positions of Earth, Mars, and the sun form an angle of 44 degrees. When this occurs, there is an opportunity (the Mars Opportunity) to launch a vehicle from Earth to Mars on a minimum-energy trajectory.

The Mars Opportunity comes once every 780

days or two years and 50 days on the average[4] during the time between planetary oppositions. Earth appears 44 degrees behind Mars in solar orbit 50 days before each opposition. Inasmuch as the ideal transfer situation is not the real one, the opportunity to fly minimum-energy trajectories on conjunction orbits come five to seven months months before opposition, as shown in table 9.1.

With advanced nuclear and projected fusion (helium 3) propulsion systems, the transfers between Earth and Mars could be cut down to three or four months. But at the time the working group was analyzing Mars missions, hydrogen–oxygen propellant appeared to be the most likely to be used for interplanetary flight.

The group was in general agreement that the exploration of Mars will require multiple missions. Some members speculated that it might take more than a century. The group recommended that in addition to establishing a permanent base, an objective of a long-term Mars program should be the development of a transportation system providing routine travel to and from Mars. This can be done by launching a space ship in a permanent solar orbit with perihelion in the vicinity of Earth and aphelion near Mars. This would allow the ship to pass each planet repeatedly and it would have to be

TABLE 9.1. Mars Conjunction Class Stopover Mission Chart

Date of Opposition	Earth Launch Date	Stopover Time (Days)	Total Mission Time (Days)
March 1997	November 1996	485	1025
April 1999	December 1998	485	1005
June 2001	January 2001	530	1020
May 2031	December 2030	500	996
June 2033	April 2033	550	950
September 2035	June 2035	530	1004
November 2037	August 2037	340	986
January 2040	September 2039	340	984
February 2042	October 2041	340	990
March 2044	November 2043	340	996

*This manifest was projected by Archie C. Young of the Johnson Space Center. The stopover times are optimized for minimum energy. See Archie C. Young, "Mars Mission Concepts." *Manned Mars Missions*, vol. I, NASA–Los Alamos, 1986.

launched only once. It would continue circulating indefinitely. Crews and supplies could then get on and off the transport in a planetary shuttle as it encountered each planet.

WATER, WATER . . .

The working group regarded the exploration and development of Mars as a long-term project, beginning with decades of pioneering. A major effort of the pioneering phase of the planet's occupation would be an assesment of water resources. This would be essential to understand the planet's thermal history and to plan for habitation.

It was expected that at latitudes less than 30 degrees, there would be little ice near the surface. At these latitudes, water is unstable and tends to sublime and diffuse into the atmosphere. Water or water ice may exist anywhere below a few hundred meters. At latitudes higher than 30 degrees, water ice may be found at depths of a few tens of centimeters, and liquid water may occur at depths of one or two kilometers (0.62–1.24 miles). The probable major reservoirs, the group surmised, would be found in the permafrost zone, the polar caps, the layered deposits near the pole, and in hydrated minerals.

With water, Mars pioneers could survive by living off the land if necessary. Water would provide hydrogen and oxygen not only for propellant but also for fuel cell batteries as a source of electricity. Oxygen extraction from the rocks or by catalytic separation from the carbon dioxide atmosphere would provide breathable air, with argon and nitrogen added to enhance the mixture.

If no trace of indigenous life is found, the working group suggested that Earth microorganisms might be introduced into the Mars environment. If the terrestrial microbes succeeded in adapting to Mars, observation of how they managed would show how human habitation could be adapted to the environment.

It was not beyond possibility that careful searching could turn up a biological oasis on the planet, an enclave where living organisms had managed to survive the environmental changes that had made Mars a desert.

A HOME AWAY FROM HOME

The search for life on Mars raises the question of whether it has ever been possible for life to exist in this cold desert. Some climatologists have speculated that Mars is locked up in an ice age that may be a cyclical stage of the climate. Discoveries that would show how Mars lost most of its early atmosphere and surface water are prime objectives of the exploration of this new world.

The cold, arid environment, with its thin atmosphere and intermittent blinding dust storms that give the planet its pink skies, poses a harsh challenge for human explorers. As shown earlier, they may expect to spend a year or more there before they can return to Earth, and longer tours are quite probable.

A minimal Mars base would include a space station–sized habitat module, a similar laboratory module, and a closed, regenerative life support system for about four to six persons. There could be a small greeanhouse for raising vegetables, a pressurized rover, tools, and a 100-k nuclear power plant.

As the interplanetary transportation system is developed with large space ships in circulating orbits, Orbital Transfer Vehicles, planetary landers, and shuttles, the Mars surface base can be enlarged. A Mars surface base occupied by 15 to 25 persons on a semipermanent basis would be the terminus of a transportation system operating on a schedule with interplanetary transports and support vehicles.

The base would be equipped not only to support surface operations such as exploration on the ground and prospecting for water and minerals, but to generate food, propellant, and other provisions for space ships as well as for the surface crew. This scenario assumed that the base can produce hydrogen and oxygen from local water sources in or below the permafrost. It would serve as the planet terminal for a shuttle operating between Mars and Phobos.

The moon, believed to have water, would be a potential source of propellant.

A conceptual study for a fairly advanced Mars base has been made by Eagle Engineering Company of Houston. It would be located on the floor of a canyon, possibly in the region of Valles Marineris, the super grand canyon of Mars that was discovered by *Mariner 9* orbiter in 1971. Initially, the base would consist of habitat, laboratory, and logistics modules of space station design; a roving vehicle; and an inflatable garage to shelter the rover from the ubiquitous Martian dust. The Eagle study depicts greenhouses on the canyon floor as the base develops to supply most of the food. The greenhouses would be environmentally controlled, filled with compressed Martian atmosphere to provide concentrated carbon dioxide for accelerated plant growth: especially fruits and vegetables. Habitat and laboratory modules of the base would be buried under a meter of soil for radiation protection.

As in nearly all other Mars base scenarios, this one would be powered by an SP 100 nuclear reactor. It would supply energy for the entire base, including an atmosphere processing plant. The plant is designed to compress the thin Martian atmosphere and convert it into a breathable mixture with addition of nitrogen and argon. Next to this unit would be the propellant manufacturing plant, which would extract and liquefy hydrogen and oxygen for propellant.

The schematic also depicts a tunneling machine on tracks. Its purpose is to excavate a second-generation base by tunneling into the canyon wall. The tunnel spoil is carried away and dumped by tele-operated hopper cars. When the tunnels are excavated, the interior walls are sprayed with a bonding resin and the tunnels pressurized with an airlock entrance.

Another study by personnel at the Johnson Space Center considered the logistics of providing the bare necessities for an operational base on the Mars surface.[5] Two configurations of the base resembling a T and lowercase b were described. They would consist of a habitat module with bunks, the environmentally controlled life support system (which reuses air and water), a laboratory module, a materials processing facility and an extravehicular activity module, with space suits and tools for working outside. The module could also serve as an emergency pressure chamber. A wash-down facility would be provided to remove dust from space suits with compressed air. One or more vehicles would be provided to move the modules or tow the laboratory to a study site or for "just moving people around the planet," the study said.

The habitation and laboratory modules would be similar in design to space station modules modified for 0.4G. A tunnel would provide safe haven against a solar flare.

The study estimated that delivery of the base elements would require two or three Mars missions with two landers per mission. Power would be supplied by the standard SP 100 nuclear power reactor. Fully developed and tested, the SP 100 reactor is the candidate of choice for power on the Moon and Mars.

T E N

CASTLES IN ORBIT

THE prospect of sending a manned expedition to Mars has been studied as a technical challenge for more than 40 years. It has evolved like our perception of the planet along with advances in space flight technology and discoveries about the nature of the planet by the Mariner and Viking missions. The state of space technology at this writing makes a manned Mars expedition feasible. As a scientific enterprise, its importance to our understanding of our native habitat, the Earth, can hardly be overstated. But like Project Apollo, such a venture can be considered realistically only as a costly national effort. Unlike Apollo, it lacks the Cold War motivation of the early 1960s that energized and funded the landings on the Moon.

An echo of that era of intraspecific competition in space may be detected in the Soviet interplanetary mission to reconnoiter the asteroidal moon Phobos, a potential Mars expedition base and source of propellant. But as the Soviet program becomes more open, competition is likely to give way to cooperative Mars ventures. Technologically, the United States and the U.S.S.R. are expected to have equal capability of developing a Mars transportation infrastructure for the 21st century based in low Earth orbit. Currently, low Earth orbit, better known by the as LEO in acronym language, is where manned interplanetary missions start in current planning. This was not always the case.

The best-known technical proposal for a Mars flight in the early years of the space age was Wernher von Braun's *Das Mars Project*, 1948, revised in 1962. As mentioned earlier, von Braun related that he made his calculations with a slide rule. In substance, he sought to show that a flight to Mars was within the capability of pre-Apollo engineering. It was his purpose, von Braun said, to demonstrate that on the basis of technologies and know how available then, the launching of a large expedition to Mars was a definite possibility. The logistics of such a voyage were no greater than those of a minor military operation extending over a limited theater of war, he said.

By 1962 von Braun had recalculated costs and logistics on the basis of technical advances since 1948. A large expedition to Mars would be possible in 15 or 20 years at a cost of only a minute fraction of the annual national defense budget, he said. Von Braun based his prediction on the development of the liquid hydrogen–liquid oxygen rocket engine, the greatest single advance in rocket propulsion, he said. In 1948, when he first plotted a Mars expedition, the only working rocket engine capable of lifting a metric ton payload was that of the V-2, and its range was only 160 miles.

Von Braun's 1962 revised expedition was planned on a large scale. It envisioned a flotilla of 10 interplanetary vessels carrying 70 men. Each space ship would be assembled in a two-hour orbit of the Earth from material delivered by three-stage rockets from the surface. (At that time, the Saturn family of rockets was being developed by the von Braun team at the Marshall Space Flight Center.) Once assembled in low Earth orbit, the flotilla of interplanetary space ships would boost themselves to Earth escape velocity and enter an elliptical orbit around the sun. The flight path was plotted so that the maximum distance of the ellipse from the sun was tangential to the orbit of Mars. At that point, the

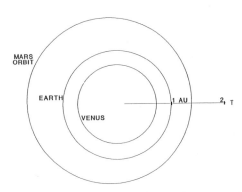

ORBITAL CONFIGURATION

Figure 10.1. The orbital range of planned planetary exploration early in the 21st century is shown on this chart. Earth's distance from the sun is charted at 1 astronomical unit (1AU). Venus is at 0.72 AU and Mars in the outer orbit at 1.52 AU. (SAIC Chart).

ships would be attracted by the gravity field of Mars. Their rocket engines would decelerate them so that they would fall into an orbit around Mars. They would stay there until their engines were fired for the return to Earth.

Three of the 10 ships would carry "landing boats" equipped with wings to descend to the surface of Mars as gliders with a landing party. One of the three landers would come down on a polar ice cap, using skis or runners, to avoid a rough landing with wheels on a rock-strewn surface. This precaution was taken to safeguard 125 tons of cargo the boat would carry, including ground vehicles. The polar landing party would use the ground vehicles for transportation to the equator to select and prepare a landing site for the other two landing craft and their crews. The polar lander would be abandoned because it could not reascend to an orbit in the ecliptic plane, where the interplanetary vessels would be parked, according to the mission plan.

Of the 70 man total crew, 50 would constitute the surface force. They would spend 400 days on the surface and then ascend to the flotilla in the two equatorial landers. The crew of 70 would leave Mars orbit for the return to Earth in seven of the 10 interplanetary ships, leaving the three that had carried cargo and the winged landers. When the returning vessels reached low Earth orbit, the crew would descend to the surface in the landing stages of the three-stage ferries that remained in LEO after supplying the materials to assemble and equip the interplanetry vessels.

The duration of this imaginative and complicated expedition was figured at two years and 239 days. The outbound voyage to Mars would take 260 days, 449 days would be spent on Mars, including the 400 days that 50 of 70 member crew would be on the surface. Transfer back to Earth would take another 260 days. "What we do not know," von Braun wrote, "is whether any man is capable of remaining bodily distant from Earth for nearly three years and returning in spiritual and bodily health."

Another unknown at that time was the density of the Martian atmosphere. Von Braun's glider landings supposed that it was 50 percent that of Earth's atmosphere, instead of .7 percent.

1963

In this period, meanwhile, an Earth orbit to Mars orbit Vehicle Design was proposed by Harry O. Ruppe of the Future Projects Office, Marshall Space Flight Center. "It can be expected that during the next decade men will travel to Mars and return safely to Earth," a document pertaining to the proposal said.[1] The proposal described an eight-man Mars mission starting in 1971 from LEO (low Earth orbit) aboard a chemically powered space ship with a mass of 2210 tons or a nuclear-powered ship with a mass of only 755 tons. Two space ships would be assembled, fueled, and launched to Mars from LEO, assuming the existence of a ground-to-orbit transport.* One ship would carry the crew and its life support system and the other would carry cargo and the Mars lander.

The entire mission would take 2.7 years, with a 450-day stay in Mars orbit. As the interplanetary transports approached Mars, they would fire retrorockets to enter a 1000-kilometer (620-mile) orbit. Crew members would descend to the surface in the lander, called the Mars Excursion Module, which would provide the habitat on the surface. Both retrorockets and air resistance (aerobraking) would be applied to control the descent.

At the completion of the surface sortie, the exploration team would return to the passenger ship in the Mars Excursion Module.† The ship would depart Mars, leaving the cargo vessel in Mars orbit. As the passenger ship approached Earth, it would enter the atmosphere by applying both rocket and air resistance braking. The crew would descend to the surface in an Earth entry module.

With nuclear propulsion, a one-year Mars mission for five men with a 10-day stay on the surface was projected for 1971. The amount of mass, including hydrogen propellant, required in low Earth orbit was 621 tons, including a 32-ton payload. The concept of a planetary lander or "excursion module" was later adopted for the Apollo lunar land-

---

* This was a 97 minute orbit, at about the orbit of Mercury or Gemini spacecraft.
† In Apollo, the lander became known as the Lunar Excursion Module or "LM".

ings. Astronauts descended to the surface of the Moon in the Lunar Excursion Module (LEM or LM), a two-stage lander. The lower stage fired rockets to brake the descent from orbit. On the surface, the descent stage became the launch pad for the ascent stage for the return to the Apollo space ship.

1966

By 1966, Mars flight theoreticians had shifted to nuclear power as the logical source of propulsion for interplanetary space ships. A nuclear-powered Mars transport was projected by Marshall Space Flight Center engineers in an influential paper read at the AIAA–AAS symposium, "Stepping Stones to Mars," in Baltimore, March 28–29, 1966.* The transportation system was based on the NERVA (Nuclear Energy for Rocket Vehicle Application) engine, which had been developed and ground tested but never flown. The nuclear stage was 54 meters (177 feet) long and 10 meters (32.8 feet) in diameter. Hydrogen was heated by the reactor and expelled through a nozzle as the propellant. The reactor was described as having a power level of 5000 megawatts thermal and weighed 309 tons. It burned 226 tons of propellant in 1775 seconds. Burning would begin at an altitude of 485 kilometers and end at 3450 kilometers accelerating the vehicle by 9 kilometers a second relative to Earth. The paper described also a nuclear–electric engine. It generated powerful electrostatic fields that accelerated charged particles (ions) through a nozzle to produce a constant thrust of 435 newtons (97.19 pounds). At that level the propulsion system was theoretically capable of driving the 388-metric-ton ship to Mars orbit in less than five months. The space ship components would be lifted to LEO by three Saturn 5 launch vehicles and assembled there. After assembly and checkout, the NERVA engine would be fired at a thrust level of 200,000 pounds to boost the ship to Earth escape velocity. Then the nuclear–electric engine would be turned on and

*The authors were Ernst Stuhlinger, J. C. King, R. D. Shelton, and G. R. Woodcock.

continue to produce thrust at a lower level to build up acceleration for the journey to Mars. The flight time to Mars was estimated at 145 days.

The transport would then enter a low Mars orbit. From there a chemically powered Mars Excursion Module would descend to the surface and later return the crew to the orbiting space ship. This scenario estimated the return trip would take 255 days after a 30-day stay on the surface. The ship would enter a high Earth orbit outside the Van Allen radiation zone. The crew would transfer to a chemically powered commuter rocket, launched from Earth to pick them up, and descend to the surface.

At the time of this proposal, the NERVA–electric propulsion system did not exist, but the technology was well known. The NERVA engine had been developed and tested by the Atomic Energy Commission. An "ion drive" that accelerated cesium ions as propellant had been developed at NASA's Lewis Research Center. It had proved effective and dependable for attitude control on a satellite.

Rising concern about the hazard of radioactivity in the event the NERVA carrier rocket failed at launch and had to be blown up by the missile range safety officer to protect a populated area persuaded NASA to postpone this development. Electric propulsion required energy on a scale that was applicable for interplanetary lift and available only from a nuclear source. Popular concern about nuclear hazards indicated that interplanetary space flight was fated to depend on chemical propulsion until it became feasible to assemble, test, and launch nuclear propulsion systems in space or on the Moon.

The Marshall engineers projected a Mars mission for the 1986 Mars opposition, when both planets were on the same side of the sun. The schedule called for Earth escape May 1, 1986; Mars capture, September 23; Mars escape, November 12; and Earth capture, July 25, 1987.

1969–1980

In 1969 the Space Task Group appointed by President Nixon projected a manned Mars mission of

640 days on the Marshall analysis of the NERVA thermodynamic nuclear rocket engine (using liquid hydrogen as propellant). Two interplanetary vessels would be assembled in low Earth orbit, each with a crew of six. They would be injected into a Mars flight trajectory March 12, 1981, and inserted into a 24 hour elliptical orbit around Mars August 9, 1982.

One half of the crew from each ship would descend to the surface in excursion modules for a period of 60 to 70 days. The landing party would then ascend to the interplanetary transports, which would blast out of Mars orbit October 28, 1982. The ships would fly by Venus February 28, 1983 to observe the planet and borrow its gravity to reshape the flight path to Earth.* Reaching Earth orbit (LEO) August 14, 1983, the crews would descend to the surface by shuttle.

The development of the reusable space shuttle during the 1970s changed the concept of manned interplanetary flight from one based on Saturn 5 technology to the one based on orbital transfer vehicles, the space station, and the shuttle.

Nuclear thermal propulsion was the preferred system for powering the orbital transfer vehicles, but that system was ecologically unusuable, as mentioned earlier. Appearing at joint hearings of the Senate Committee on Aeronautical and Space Sciences and the Joint Committee on Atomic Energy February 23, 1971, George M. Low, acting NASA administrator, characterized the NERVA engine as a "space propulsion system to be used only in Earth orbit on out." It needed the shuttle to put it into orbit, he said.

The development of nuclear–thermal and nuclear–electric propulsion was halted in 1972 and chemical propulsion was advanced by uprating the Saturn hydrogen–oxygen engines for the shuttle main engine system. It took nearly a decade to build the new engine and make it reasonably reliable after repeated failures of the pressure pumps and other components. Still, with peak specific impulse of 465 seconds, the shuttle hydrogen–oxy-

gen engine was only half as efficient as the NERVA engine (825 seconds) which potentially could deliver twice the payload to a destination in space in half the time of any chemical engine.

Hydrogen–oxygen propulsion thus became the standard for interplanetary mission planning for the early 21st century. By 1984 a class of Orbital Transfer Vehicles (OTVs) was considered the next step beyond the shuttle to operate from the space station in LEO to geosynchronous orbit (GEO) and lunar orbit.

When this system of interorbital transfer was in place, it was theoretically feasible to launch manned expeditions anywhere in the Solar System, once the required mass (propellant and vehicles) was established in LEO. In September 1984, the Science Applications International Corporation, a private consultant, projected such a system to carry out a manned mission to Mars for the Planetary Society, an organization of scientists and laymen in Pasadena, California. The consultant drafted a plan for a three-year mission to Mars with a crew of four. Three would descend to the surface for 30 days while one remained in orbit in the command module of the Earth to Mars transfer vehicle.

Two interplanetary transfer ships would be launched on the mission, one for the outbound flight and the other for the return to Earth. Each would be launched from LEO by a train of Orbital Transfer Vehicles thrusting in sequence to inject them into a hyperbolic return trajectory. In this flight path the transfer ships would swing around Mars at different times and head back to Earth.

The entire mass of vehicles, including 10 Orbital Transfer Vehicles (OTVs), would be lifted to Earth departure orbit by 18 shuttle launches, each shuttle hauling 60,000 pounds of cargo. The outbound transfer ship would be launched June 15, 2003, with the Mars lander and orbiter from the space station at 230 mi altitude by a stack of seven OTVs. Each would boost the transfer ship into a sequence of increasingly elliptical orbits by firing engines at perigee.† At the seventh boost the outbound ship would be injected into an Earth escape, hyperbolic Mars return orbit.

* A Venus swing-by takes advantage of Venusian gravity to accelerate or decelerate the velocity of the passing vehicle. Acceleration is advantageous on the outbound leg of the Mars journey; deceleration is the advantage on the inbound leg. In each case, the planet's gravitational assist saves propellant.

† Energy expended at perigee, the low point of an Earth orbit, increases the distance of apogee, the high point.

The outbound Mars Mission ship would consist of three sections: (1) the interplanetary ship with a logistics module, solar arrays for power, radiators to eject surplus heat, a pressurized habitat for the crew of four, an experiment module, an extravehicular activity station, and a docking module, all connected by a tunnel; (2) the Mars orbiter, with solar arrays, radiators, a habitat/experiment module, the Mars Departure Vehicle, and high-gain antenna to communicate with Earth; (3) the Mars lander, with descent and ascent stages and an aerobraking device 54 meters (177 feet) in diameter to slow the descent from orbit.

The Earth return ship with less mass than the Mars Mission ship, would be accelerated Marsward by three OTVs in a six-hour launch sequence. It would be launched 10 days ahead of the Mars Mission ship (June 5, 2003) but would arrive a month later, at the end of the 30-day exploration period. At that time the landing party would have returned to the Mars orbiter in the lander ascent stage. The crew would then transfer to the Earth Return Vehicle as it passed around Mars by making rendezvous with it in the Mars Departure Vehicle, which is part of the orbiter. The Earth Return Vehicle would carry an aerobrake 13.3 meters (43.6 feet) across.

Following its launch from LEO June 15, 2003, the outbound ship with a mass of 265,320 pounds would arrive at Mars December 24, 2003. The Earth Return Vehicle with a mass of 94,600 pounds would arrive to pick up the crew January 23, 2004. The crew would arrive at Earth orbit June 5, 2006. In the event the Earth Return Vehicle failed on the journey to Mars, the Mars landing would be aborted and the crew would remain aboard the outbound ship as it swings around the planet and goes back to Earth. This would require a powered swing-by maneuver.

An array of other contingencies was considered by the consultant, including a failure of the lander and orbiter. That would require retargeting the outbound vehicle fly-by of Mars. The vehicle would have the capability of making a velocity change of only 0.7 kilometers a second and in order to accomplish the retargeting maneuver, the crew would have to jettison 23,000 kilograms of mass.

The major goal of the surface team would be to select a site for a future Mars base. The team would make photographic surveys, deploy meteorological instruments, set up a mobile geophysics laboratory, and collect 550 pounds of surface rocks and dirt to be returned to Earth for analysis. During the interplanetary cruises, the crew would make solar wind, solar flare, gamma ray, and ultraviolet telescope observations.

AEROBRAKING

The NASA working group report on manned Mars missions considered the technique of aerobraking as a significant design requirement for interplanetary transfer ships. As mentioned earlier, the technique dissipates the kinetic energy of the interplanetary segment of the flight by passing the vehicle through a planetary atmosphere to slow it down so that it falls into orbit around the planet. Aerobraking has been used to reduce the entry velocity of manned spacecraft since the beginning of space flight.[*] The U.S. space shuttle was the first winged glider to use it successfully. In 1976 the *Viking* landers were the first interplanetary vehicles to use it. On arrival at Mars, each was deployed from its orbiter enclosed in an aeroshell, a heat shield that served as an aerobrake as the the lander entered the thin Martian atmosphere at 151 miles altitude. The aeroshell on each lander was 11.5 feet in diameter. Its drag reduced descent velocity until it was jettisoned by springs at 19,000 feet when parachutes were deployed. Terminal descent engines ignited at 4600 feet, reducing descent speed to about 5 miles an hour. At touchdown, switches on the lander's extended legs cut the engines off.

The working group reported that for conjunction class missions to Mars, the departure mass of a vehicle equipped for aerobraking will be 20–30 percent less than the mass of a vehicle, depending on propulsion for braking into orbit. For opposition class missions, aerobraking may save as much as

---

[*] Mercury, Gemini, and Apollo capsules made ballistic reentries into the atmosphere, where their kinetic energy was transformed into heat. At lower altitudes, parachutes were deployed to complete the descent. Entry vehicles using aerobraking are heavily heat shielded.

60 percent of departure mass (propellant) as these flights encounter a planetary target at higher velocities. Although aerobraking structures make vehicle design more complex, the group said, they may allow an interplanetary ship to delete an entire propulsive stage that would be necessary for an all-propulsive mission.

The working group presented a description of an all-propulsive space transport that would use propulsive capture at Mars and Earth for a mission at the 1999 opposition. The propellant elected was hydrogen–oxygen. The group noted that a propulsive mission using storable propellant was considered, but its higher weight made it impractical.

For the all-propulsive interplanetary vehicle launched from low Earth orbit, three propulsive stages are necessary. The vehicle is 92 meters (301.8 feet) long. It carries a Mars excursion module with a combined propulsive and aerobraking system for the descent to the surface from Mars orbit.

The first stage, consisting of three tanks containing 1,026,602 kilograms of propellant, used shuttle-derived tanks and main engine to propel the vehicle out of Earth orbit. The Earth departure stage has a mass of 1,083,127 kilograms. The second and third stages use propulsion systems derived from Orbital Transfer Vehicle technology.

The second stage is fired once at Mars encounter to drop the vessel into Mars orbit and again to depart Mars orbit and return to Earth orbit. The third stage is fired for braking into Earth orbit. The forward section of the vehicle is the mission module to which the Mars excursion module is attached. In addition to descent and ascent engines, the excursion module carries a 15.2-meter (49.86-foot) aeroshell, which functions as an aerobrake.

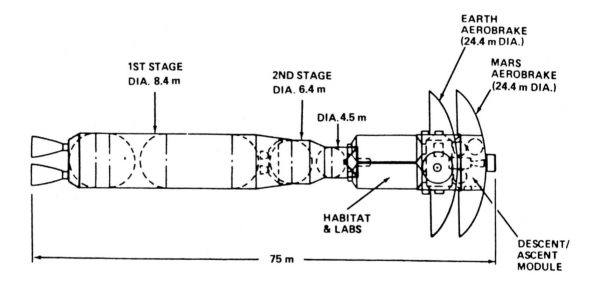

	PROPELLANT WEIGHT* (KG)	TOTAL WEIGHT* (KG)
1ST STAGE	445,359	484,691
2ND STAGE	72,474	78,951
SPACECRAFT	32,661	149,729
TOTAL	550,494	713,371

* REFERENCED TO EARTH

Figure 10.2. One 21st century option is a manned Mars mission. In this sketch, a Mars transport is depicted with two aerobraking systems to reduce interplanetary flight velocity for capture in Mars orbit and again capture at Earth orbit on the return. The Mars lander would use a combination aerobraking and retrorocket descent system. (NASA-Los Alamos National Laboratories.)

Remaining in orbit at Mars, the mission module consists of habitat, laboratory, and logistics modules similar in architecture to those designed for the space station. The similarity in outer dimensions is dictated by the the size of the 60 by 15 foot shuttle cargo bay in which the components of the interplanetary vehicle are lifted to Earth orbit. Thus, the Mars vehicle modules might be sized at 42 to 44 feet long and 14 feet in diameter.

The all-propulsive vehicle has a total mass of 1,623,298 kilograms at launch from LEO. The use of aerobraking at Mars and Earth would allow a significant reduction in mass and size of the interplanetary vehicle.

Equipped for aerobraking, the Mars vehicle would have a total mass of 713,371 kilograms, less than one half that of the all-propulsive vehicle, including propellant mass of 550,494 kilograms, a little more than one third of the propellant required from the all-propulsive option. Only two propulsive stages are required, one for Earth orbit departure and the other for Mars orbit departure. Each stage is smaller than those on the all-propulsive vehicle. The vehicle would carry two aerobrake shells, each 80 feet in diameter, one for aerocapture at Mars and the other for aerocapture at Earth. Presumably, the Mars aeroshell would be used also to brake descent to the surface. There is a suggestion in the report that the aeroshell might be used on the surface for storage or habitation.

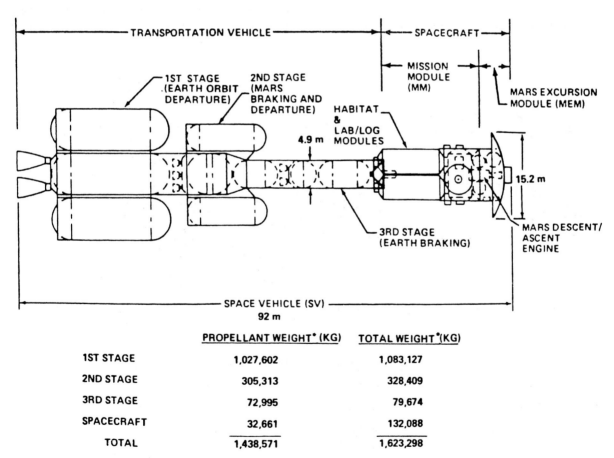

	PROPELLANT WEIGHT* (KG)	TOTAL WEIGHT *(KG)
1ST STAGE	1,027,602	1,083,127
2ND STAGE	305,313	328,409
3RD STAGE	72,995	79,674
SPACECRAFT	32,661	132,088
TOTAL	1,438,571	1,623,298

* REFERENCED TO EARTH

MMM 1999 opposition all-propulsive option.

Figure 10.3. In this Mars mission option, capture at Mars and return to Earth orbit is all propulsive. Economically, the volume of propellant required for braking at Mars and at Earth is considered impractical. Only the Mars lander would use a combination of aerobraking and propulsive braking. The manned Mars misison could be flown at the 1999 Mars opposition. (NASA-Los Alamos National Laboratories)

Decelerating the interplanetary vehicle from hyperbolic velocity on approach to a planet to orbital velocity by aerocapture requires guidance through an entry corridor in the planet's atmosphere. The angle of attack must be precise; otherwise the vehicle skips out of the atmosphere before shedding hyperbolic velocity, like a flat stone skipping across the surface of a pond. Trajectory analyses of Earth to Mars transfers from 1999 to 2028 calculated the velocity at which the interplanetary vehicle would enter the atmosphere of Mars at 17,700 to 30,000 feet a second. The entry velocities are higher on opposition class missions than on conjunction class missions.

On the return to Earth the maximum entry velocity for conjunction class missions was calculated at 38,000 feet a second. Opposition class entry velocities were expected to be higher. The working group warned of a hazard of using aerobraking at Earth to reduce hyperbolic velocity on the return from Mars. The resulting deceleration by Earth's atmosphere may impose excessive forces on human passengers, more than they can withstand without physiological damage. Where deceleration forces appear threatening, propulsive entry should be substituted.

ARTIFICIAL GRAVITY

The technology required for manned Mars expeditions has been developed except for the Orbital Transfer Vehicle, which is essentially an extension of existing upper-stage designs. But one question remains unanswered. That is how humans will fare in long term exposure to microgravity on interplanetary voyages? Will artificial gravity be necessary? As indicated earlier, the life sciences people in NASA who are studying the effects of microgravity on human muscle, bone, and heart don't know. The Soviet physiologists point out that their space station experience (1988) shows that a cosmonaut does recover from a year's exposure to microgravity in a space station on return to Earth. But no data yet exist showing how a crew person would fare on a planet with two-fifths G after such exposure. Could a cosmonaut or astronaut function on two-fifths G Mars after a one-year microgravity trip from Earth,

even if he exercised every waking moment during the trip? The working group studied two artificial gravity options, 1 g and .4 g.

In the 1-g option design the habitat and laboratory modules would be mounted at the ends of a 400-foot transverse truss consisting of 200-foot outriggers on each side of the vehicle just aft of the Mars excursion module and aeroshell. It would be rotated by small thrusters at four revolutions a minute to produce a force in the modules of one gravity. The report said that the physiological tolerance of crew persons to coriolis force (produced by a rotating body) was limited to two to four revolutions a minute. Astronauts moving from a module at one end of the outriggers truss structure through a tunnel to a module at the other end would experience variations in gravity with distance from the spinning central core, the Manned Mars Missions study noted.[2] The central core would contain a shielded, despun command module serving as a command station and shielded storm shelter for the crew during solar flares, according to an Eagle engineering design.[3]

In its summary of vehicle concepts, the working group depicted the vehicle equipped with artificial gravity outriggers as all-propulsive. On a vehicle with aerobraking, some of the modules would have to be relocated behind the aeroshell. Other design and operational difficulties were noted by the summary. Because of the 400-foot separation of the habitat and laboratory modules, efficient use of the total habitation environment is compromised. Frequent visits from one module to the other through the long tunnel with variable g forces would tend to result in sickness. Some systems and living quarters would have micro-g, partial g and 1 g environments, with g forces acting in different directions.

The working group examined another artifical gravity option that would provide 1.4 g. Here the rotating boom would be 36.6 meters (120 ft) long with modules at the end of each 18.3-meter (60-foot) outrigger. The group analysis said that although the distance to be traversed between the modules would be less than on the 1-g option, the same difficulties would exist, although to a lesser degree.

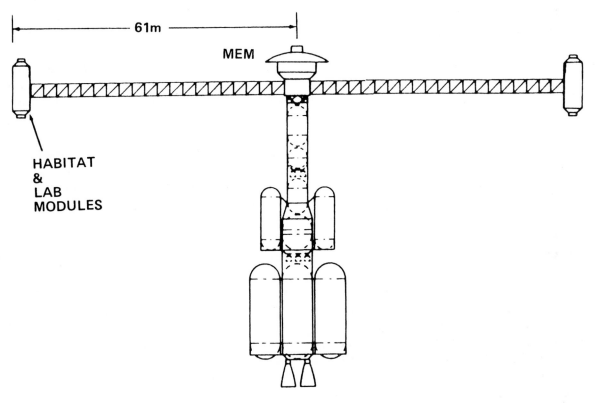

MMM 1-g option.

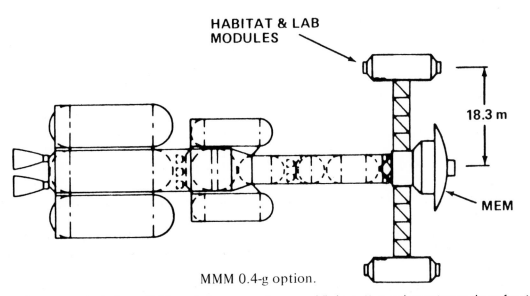

MMM 0.4-g option.

Figure 10.4. The question of whether artificial gravity is necessary for manned flights to Mars and operations on the surface is far from settled. Two levels of artificial gravity can be provided in flight by rotating habitat modules at the ends of outrigger tunnels. The tunnels extend perpendicular to the central core of the interplanetary space ship or station which contains the propulsion unit. A Mars Mission Working Group reported that human physiological response to coriolis force requires that the rate of rotation be limited to 2 to 4 revolutions per minute. Fig. 1 shows an artificial g system providing a 1 g field for modules at the ends of 200 foot outriggers rotating at 4 rpm. Fig. 2 shows a system with 60 foot outriggers that provides 0.4 g at the same rate. (Working Group Report, Manned Mars Missions. NASA-Los Alamos National Laboratories, Revision A, Sept. 1986)

CIRCULATING ORBITS: THE 15 YEAR CYCLE

Although a manned Mars expedition appears to be well within the technological capability of the United States at this writing, any exploration program requiring repeated visits faces the prospect of prohibitively high cost. Most of it is the charge for energy to lift a space ship from an orbit at 1 astronomical unit (AU) to an orbit of 1.52 astronomical units via minimum-energy conjunction class or higher-energy opposition class trajectories.

Suppose the transfer from 1 to 1.52 AU could be made without expending energy. It could be done by a space ship in a circulating orbit around the sun passing near Earth at one point and Mars at another, with only minor course adjustments. This is a third class of trajectories. It has been known for more than twenty years.[4]

Circulating orbits have been considered in the context of a fifteen-year cycle that characterizes the geometrical relationship between Earth and Mars. The relationship is shaped by the fact that the orbit of Mars is highly eccentric, exceeded only by the orbits of Mercury and Pluto. At perihelion, Mars is 128.41 million miles from the sun, or 1.38 AU; at aphelion Mars is 1.66 AU, or 154.81 million miles from the sun. The figure 1.52 AU is the average. The difference of 26.45 million miles not only has profound climate implications for Mars, as mentioned earlier, but is a factor in determining flight schedules. By contrast, Earth's orbit around the sun is nearly circular, ranging only from 91.34 to 94.56 million miles. The mean value is 1 AU.

Because the eccentricity of the Mars orbit is so much greater than that of the Earth orbit, the distance between the two planets varies from one opposition to the next. It ranges from 35 to 63 million miles approximately. The most favorable opposition when the planets are 35 to 36 million miles apart occurs every 15 or 17 years.

As a means of defining the way circulating orbits would work, analysts have approximated the orbital period (synodic) of Mars at 1.875 Earth years.[5] Every 15 years as Earth makes 15 revolutions around the sun, Mars makes 8. Oppositions occur on the average every 780 days (or about every 26 months).

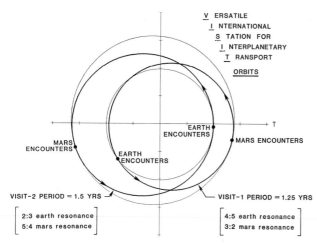

Figure 10.5. The feasibility of establishing a long term or permanent base on Mars is enhanced by the VISIT orbits concept. The term is an acronym for Versatile International Station for Interplanetary Transportation. In a VISIT orbit around the sun, the station or space ship would encounter Mars and Earth at known intervals. Planetary access would be provided by local transports making rendezvous with the interplanetary station. With relatively minor course adjustments, the planetary encounters would continue indefinitely. This chart parepared by Science Applications International Corporation depicts two types of VISIT orbits. VISIT I with a period fo 1.25 years would encounter Mars every 3.75 years and Earth every 5 years. VISIT II with a period of 1.5 years would encounter Mars every 7.5 years and Earth every 3 years. (SAIC)

It takes between seven and eight oppositions for the planets to resume their relative orbital positions, so that during every seventh or eighth opposition they will repeat the same geometrical relationship.

Seven oppositions cover about 15 years, and eight take about 17 years. On the 15 year cycle, the geometry is repeated every 2.1429 years. As a consequence, trajectories connecting Earth and Mars will occur every 2.1429 years. Space ships in circulating orbits could depart Earth at that interval. In a circulating orbit with perihelion at 1 AU, passengers would board the interplanetary space ship in the vicinity of Earth and transfer to a Mars orbit shuttle at aphelion, approximately at 1.52 AU. Once established in the circulating orbit, the interplanetary ship would continue to pass these transfer points indefinitely, with minor course corrections for drift, the analysis said. Propulsive energy would be used only once to establish the vessel in it circulating orbit.

Under the influence of the gravitational field of the sun, like a comet, the ship would continue to circulate among the two planets. The analysis added that even the energy for minor course corrections was free. It would be provided by gravity assist from the Earth and Mars.

Circulating orbit transfers offer a solution to the problem of supporting a long-duration base on Mars from Earth, or rotating crews at a Mars base and establishing trade with Earth and the Moon. Two types of circulating orbits have been examined. One is called the VISIT orbit (Versatile International Station for Interplanetary Transportation).[6] It was designed for low relative velocities as the circulating space ship passed Earth and Mars terminals. In order to achieve this, the perihelion of the transfer orbit must be at 1 AU, where the Earth, is and the aphelion at 1.52 AU, where Mars can be intercepted. An orbit with these characteristics has a period of 1.25 years. It completes four revolutions of the sun while Earth completes five; and three while Mars completes two. Earth will be encountered every five years and Mars every 3.75 years.

The VISIT orbit was proposed by John Niehoff and Alan L. Friedlander of Science. It was designed from a group of circulating orbits that had

low velocities at Earth and Mars encounters, facilitating transfer to the planets. The sun-circling track had to be nearly tangential to the orbits of the planets to enable transfer to be made with minimum energy. That required the circulating space ship to come nearest the sun at 1 AU on the Earth pass and reach aphelion at 1.38–1.66 AU at Mars passage. A second VISIT orbit (VISIT 2) was charted with a period of 1.5 years. It would complete two revolutions around the sun while Earth completes three, and five revolutions while Mars completes four. Earth encounter would occur every three years and Mars encounter every 7.5 years.[7]

Science Applications International analysts said that the eccentricity of the Mars orbit makes the VISIT 1 orbit ideal for encountering Mars near its perihelion and the VISIT 2 orbit ideal for Mars encounter at its aphelion.

Another type of circulating orbit is the Up/Down Escalator Orbit, first proposed by Edwin E. (Buzz) Aldrin, Jr., who landed on the Moon in 1969 with Neil Armstrong in *Apollo 11*. It offers regular and more frequent encounters with Earth and Mars that come sequentially about every 2 1/7 years. Encounter speeds would be higher than those of the VISIT orbits and require more energy to make the transfer to the planet. The Up escalator orbit (from Earth to Mars) would have a short transfer period to Mars and a long transfer back to Earth. The Down escalator would be the reverse—short to Earth, long to Mars.

Castles and Taxis. The prospect of the VISIT and Up/Down Escalator orbits offered a potentially economical means of sustaining a long-term base on Mars. A detailed description of these transportation modes envisioned a large space liner injected into a circulating solar orbit for periodic encounters with Earth and Mars, with all the facilities of a space station. The ship was called a CASTLE, an acronym for Cycling Astronautical Spaceship for Transplanetary Long Duration Excursion.[8]

At each planet encounter, passengers, equipment, and provisions would be transferred from the CASTLE to a space station by a local shuttle called a Taxi. The Taxi would bring the Mars base crew

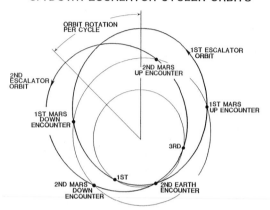

UP/DOWN ESCALATOR CYCLER ORBITS

Figure 10.6. The Up-Down Escalator orbit proposed by Aldrin offers encounters with Mars and Earth sequentially every 2 and ½ years. Encounter speeds would be higher than those of VISIT orbits. (SAIC)

from the station to the CASTLE for their return to Earth. Shuttles would provide transportation between the surface and the orbital stations at Earth and Mars. Taxis would make the rendezvous and dock with the CASTLE on an outbound leg as it passed by.

The CASTLE would provide many of the facilities of a well-equipped base with .25 artificial gravity during a transit of six to 30 months.[9] It would be assembled in Earth orbit and have the dimensions of an ocean liner, with capacity for a large tonnage of provisions workshops, laboratories, and recreation facilities. Transfers from a CASTLE in a circulating orbit might be compared with conventional air travel in which the passengers transfer from the airplane at an airport to a taxi or bus. The interplanetary spaceport is a space station in planetary orbit. At Earth it might be located in low Earth orbit, in geostationary orbit, or at an Earth–Moon or Earth–sun libration point.

A basic CASTLE was visualized in this scenario as having a dry mass of 880,000 lb with a crew of six to 21 persons. On conjunction flights, it would require a propulsion system for capture at Mars and Earth, but its principal voyages would be made in circulating orbits, where a large propulsion system is not needed. However, maintenance facilities would be required for the Taxis, including a pressurized hangar with a mass of 132,000 pounds. It would bring the total mass of the CASTLE to more than a million pounds.

In the VISIT or Escalator orbits, maintenance and overhauls of the CASTLE must be performed in transit, the scenario noted. (The alternative would be to deflect the ship into a planeary orbit and relaunch it.) It was estimated that about 35 percent of the basic mass of this huge vehicle must be delivered to it.

As crews come and go at each planetary passage, the CASTLE must have an operating crew cadre at all times. It is always manned, despite advanced automation and robotics. The scenario estimated that a standby or maintenance crew of six persons working three shifts with two per shift would be required at minimum. During Earth and Mars passage, two crew members may be enough to work the ship while others have shore leave.

(However, a sudden transfer from 0.25 G to 1 G would likely create physiological problems for those taking shore leave.)

One of the most active and authoritative think tanks projecting advanced space flight concepts, Eagle Engineering Company of Houston refers to CASTLES as Periodic Space Stations (PSS).[10] These are space stations with essentially the same architecture as the permanently manned station NASA designed for the mid-1990s. The difference is that the PSS is equipped with a powerful propulsion system that boosts it out of low Earth orbit into a solar orbit with Mars at aphelion and Earth at perihelion. Thus, the space station is transformed into a cycling space ship.

Eagle presented the PSS as a hotel in space for up to 20 persons with a mass of 240–350 metric tons. It would be equipped to provide variable gravity, so that a crew leaving Earth could be acclimated to the 0.38 g of Mars. Initially, Eagle suggested, the PSS could be used as a research station in low Earth orbit; conversely, a space station developed for Earth orbit could become the model for the PSS.

The cycling space ship would be loaded with 10 metric tons of supplies at Earth passage by a transfer stage. The stage would consist of a 5-ton Manned Mission Module and a Logistics Module hitched to two 40-foot orbital Transfer Vehicles in series. One OTV would act as the first-stage booster. When its propellant is exhausted, it would return to Earth orbit by aerobraking. Then the second OTV with the crew (Manned Mission) module would make rendezvous and dock with the Mars-bound PSS. It would go along to Mars with enough propellant for Mars arrival operations while the crew would transfer to "hotellike accommodations" of the PSS.

These accommodations would consist of habitat, logistics, laboratory, observation, power, and propulsion modules connected by 5-ft-diameter booms. Eight space station modules would be assembled to make up the "standard" PSS. Four modules would be located on outrigger booms extending 735 ft from the central axis of the station, two at each end. The 735 foot radius of each of the two outriggers was calculated to provide 1 g at two revolutions a minute.

Figure 10.7. Artist's sketch diagrams a circulating orbit linking a planet, such as Mars, with its moon, such as Phobos. (NASA Pathfinder)

Three modules are attached to the despun axis of the station, two designed as food-growing greenhouses in microgravity with large skylights that can be opened to sunlight. A third microgravity module is located near the hangar and propellant storage area and is used as a workshop and control center for the hangar and docking system. A radiation shelter is also located in the 1-g area, and a health maintenance and recreation module is provided between the modules.

When the PSS reaches the vicinity of Mars orbit the Mars-bound crew enters the Manned Mission Module attached to the Orbital Transfer Vehicle and leaves the station. The OTV maneuvers to an aero-intercept of Mars to reduce velocity by aerobraking and to enter an elliptical orbit of Mars with apoapsis near the moon Phobos. The OTV now can make rendezvous at a Mars space station which is in orbit near Phobos, a source of propellant for the Earth–Mars run. The Phobos station is similar in design to the PSS and serves as a propellant storage depot and transfer node to the Mars surface base. At Phobos station the crew enters a reusable lander to descend to the surface.

To return to Earth, a crew must wait for a PSS to cycle to the short Mars–Earth segment of the escalator orbit. Crew and supplies are then loaded aboard the Mars Mission Module and boosted to the PSS by the same OTV that brought the crew to the flying hotel from Earth. The OTV meantime has been refueled at Phobos station. On the Mars–Earth leg of the Down Escalator orbit, return to Earth would take about six months. This is the fast lane of that circulating orbit.

The Base. The main assumption underlying the concept of the CASTLES or Periodic Space Stations is the existence of an operating base on Mars. Science Applications and Jet Propulsion Laboratory analysts predicted that a mature base would be developed in the 2035 time period.[11] Corollary assumptions held that cryogenic (oxygen–hydrogen) propulsion systems would be used for CASTLES and Taxis and that extraterrestrial propellant would be available. The propellants would be manufactured on the Moon, Mars, and Phobos, in addition to Earth itself.

A Mars base projected by Eagle Engineering is illustrated as established in "an ancient, water-eroded canyon."[12] Explorers wear pressure suits designed by NASA's Ames Research Laboratory. They travel on the surface in a specialized traverse vehicle with guidance, navigation, and communication equipment. It tows a habitat and a logistics module.

The initial base projected by Eagle and the Johnson Space Center would consist of habitat and laboratory modules of space station design, an extravehicular activity module for donning and removing space suits, a roving vehicle, and an inflatable garage. The function of the garage is to keep Mars dust off the vehicle (or vehicles). The extravehicular activity module would be equipped with a compressor to blow Martian air on space suits to remove dust.[13]

The advanced base described by Eagle spreads over the canyon floor and includes tunnels bored in the canyon wall for safe havens from solar flares. A greenhouse on the canyon floor would supply most of the food for the base. It would consist of environmentally enclosed structures in which compressed Martian atmosphere would provide concentrated carbon dioxide for plant growth.

In the advanced base, the habitat and laboratory modules would be under three feet or more of soil, for radiation (cosmic) protection. Equipment would include a lightweight crane and trailer, which would be used for moving base components, such as the modules; an SP 100 nuclear power reactor; the greenhouse structures; and an atmosphere processing unit. The processor would be designed to compress the Martian atmosphere and convert it to a breathable atmosphere for humans. Another processing facility would be set up to manufacture propellant from the atmosphere and from water. Equipment would include the tunneling machine. When excavated, tunnels would be sprayed with a bonding resin and pressurized. A pilotless aircraft would be provided for photo reconnaissance flights and scientific observations.

In order to sustain the Mars base continuously, three transportation options were examined by analysts: conjunction, VISIT, and escalator orbits. If the conjunction mode of interplanetary transfer were used, two interplanetary CASTLES would be required for direct flights. Each would be launched from Earth orbit at 4.3-year intervals with a crew of 19. They would return on the second CASTLE after 3.2 years at Mars (on the average). Their total tour would be 4.8 years.

The turnaround time at the Earth terminal for refurbishment and refueling would average 1.6 years before each vehicle is launched again. Together, the two transports would make seven flights to Mars in a period of 15 years. During that cycle, the first CASTLE would make four round trips and the second would make three. The first ship would be in orbit around Mars by the time of the next opportunity to launch from earth.

The analysts assumed that by the time a Mars base is operating, the space transportation infrastructure would include an Earth spaceport at the L-1 libration point in cislunar space.[*] The noncirculating conjunction mode flight would start with the concentration of propellant at the L-1 spaceport. Three low-orbit tankers would be equipped with crew modules to transport the crew and the consumables it requires to L-1 before launch.

The conjunction flight is all-propulsive, with aerocapture at Mars and Earth. The CASTLE's tanks must carry sufficient propellant for the capture burn at Mars, a maneuver omitted by a vessel in circulating, nonstop solar orbit. This option requires a stopover in Mars orbit and additional propellant compared to the VISIT mode of the circulating orbit option. For the launch from L-1 to Mars, four Earth orbit tankers are required to fill the propellant tanks.

For the trip to Mars, the artificial gravity section of the CASTLE is spun up as the vessel reaches escape velocity. It must be spun down before capture at Mars. The CASTLE then enters an elliptical orbit around Mars, where, the analysts believe, natural planetary perturbations will align the orbit for departure to Earth.[14]

Propellant for the return trip is loaded from the Mars spaceport by Orbital Transfer Vehicles used as tankers. A Mars shuttle delivers the homebound crew and its provisions to the CASTLE. With boost assist from the OTVs, the CASTLE departs Mars orbit and the artificial gravity section begins spinning. At the approach to Earth, the spin is stopped. The CASTLE's propulsion system provides the capture burn which enables the big ship to deliver the crew to the L-1 spaceport. From there, the crew is ferried to the low Earth orbit space station aboard tankers with personnel modules.

The second flight option is the VISIT orbit mode with CASTLES in a circulating orbit. The VISIT orbit mode can sustain the Mars base with three CASTLES. Each would carry 21 persons, 15 for a tour of duty on Mars. In a 15-year period, eight transfers can be made from Earth to Mars. Various crews would have tours varying from 5.7 to 7.9 years, with stays at Mars ranging from 1.6 to 5.9 years.

The third option is the Up/Down Escalator class of orbit transfers. A single Up and a single Down Escalator CASTLE would maintain the size of the Mars surface crew within guidelines set by the space agency. The use of the Up orbit for Earth to Mars transfer and the Down orbit for return would minimize transit time. Each CASTLE would make seven round trips in the 15-year cycle with a crew in transit either to or from Mars on each trip.[†] The sequence of events for the Escalator trips is the same as that of the VISIT orbit trips. However, the Escalator orbits require higher escape velocities at both Earth and Mars and that imposes increased propulsion by the Taxis carrying passengers and supplies to the CASTLE at Earth and Mars passages.

At a spaceport in Earth orbit, preferably at the L-1 libration point in cislunar space, the augmentation is provided by auxiliary propellant tanks which are jettisoned before the Taxi docks with the CASTLE. At the Mars spaceport, possibly in orbit near Phobos, an Orbital Transfer Vehicle would supplement Taxi boost at every departure. In some instances, said the analysts, auxiliary propellant tanks would be needed to meet higher velocity

---

[*] The L-1 libration point is one of five locations of gravitational equilibrium in the Earth–Moon system. A structure "anchored" there would remain indefinitely in a stable position relative to the Earth and the Moon.

[†] Because the orbit of Mars is more eccentric than that of Earth, the distance between the planets varies from one opposition to the next over a 15-year period, as detailed earlier. The variation repeats every seven to eight oppositions.

needed to catch up to the CASTLE as it rounds aphelion and heads back toward Earth orbit.

Occasionally, the Escalator orbits require propulsive maneuvers by the CASTLES to maintain favorable orbital alignment within the Earth–Mars system. This requirement is in addition to realignment required by gravitational effects on the orbit each time the CASTLE passes Earth or Mars.

Although the sun's gravitational field provides most of the energy for circulating orbit transportation, it does not add up to a free lunch. Humans have to provide some of the energy in the form of liquid hydrogen and liquid oxygen.

The Science Applications–Jet Propulsion Laboratory analysis of requirements for sustaining a Mars base calculated that the Up/Down Escalator orbit would consume 32,830 metric tons of hydrogen–oxygen propellant over a 15-year period while the VISIT orbit would consume only 21,500 tons. Flying a sequence of conjunction stop-over orbits during that period would take 30,205 tons. The analysis also showed that only about 10 percent of the fuel —liquid hydrogen—would be obtained from Earth. Mars and Phobos would provide the balance of the liquid hydrogen and all of the liquid oxygen. Most of the propellant would be obtained from the Mars system.

A revision of the Escalator orbit system would make it considerably more economical in terms of propellant usage. By eliminating the Up Escalator and using the Down Escalator for the transfer to Mars as well as the transfer back to Earth, propellant usage could be reduced to 23,660 tons for the entire cycle, the analysts calculated. Only one CASTLE would be required for this option. However, its disadvantage is longer travel time on the outbound leg to Mars. The Down Escalator return leg is the shortest of all the modes.

IN SUM

When the circulating orbits were initially analyzed, it appeared that these modes would use less propellant than the conjunction stop-over mode because they would not require repeated boosts of the CASTLES to and from Mars. One launch from the vicinity of Earth would put them on the Earth–Mars–Earth track indefinitely. But after that one launch, there are still propellant requirements.

All the circulating orbits, the analysis found, turn out to have higher relative velocities at planets than the minimum-energy conjunction orbits. That requires the Taxis to reach higher velocities than needed for the conjunction mode to make the rendezvous with the CASTLE as it passes the planet. This factor plus the need for the Taxis to transport consumables to the CASTLE may result in the use of as much or nearly as much propellant for a circulating orbit mode as for the conjunction mode. In this analysis, hydrogen–oxygen propellant was considered in a state of the art cryogenic propulsion system. Nuclear electric or nuclear thermodynamic propulsion would change this picture.

The VISIT orbit mode with low planet relative velocities would require lower Taxi velocities for making rendezvous with a passing CASTLE. In terms of propellant, there would be a saving in the Taxi "fare." It is the most conservative in propellant use of the Earth–Mars transportation options, but it requires three CASTLES instead of one or two for the Mars transit system.

These scenarios are the preliminary projections of a manned interplanetary transportation system for the 21st century. The technology to create it exists now. Much of the basic engineering is encompassed in the shuttle and will be developed in the space station.

It has been nearly a century since the Russian schoolteacher Konstantin Tsiolkovsky uttered his famous dictum: "The Earth is the cradle of mankind, but one cannot forever remain in the cradle." The prophecy the dictum implies is now being fulfilled with the technical evolution of space-faring societies. For the last 20 years, the permanent manned space station has been cited as the next logical step for the United States. It is the next, but not the last, for it opens the door.

EPILOGUE

I N mid November 1989, NASA submitted a report of a 90-day study of Human exploration of the Moon and Mars to the National Space Council. The study was a response to the call by President George Bush July 20, 1989 for "a long range, continuing commitment" to human expansion into the Solar System. It examined ways by which this could be done in the early 21st century.

The course charted by the President was clear and direct: "First, for the coming decade, for the 1990s, Space Station Freedom, our critical next step in all our space endeavors," he said. "And next, for the next century, back to the Moon, back to the future, and this time back to stay. And then a journey into tomorrow, a journey to another planet, a manned mission to Mars. Each mission should and will lay the groundwork for the next."

Thus for the second time in American history, a new space initiative was proclaimed by the nation's president. But now with the passage of 28 years since John F. Kennedy called for a manned landing on the Moon, the new initiative arises out of a greatly altered political and technological background.

The Cold War which had largely motivated the lunar adventure in the Kennedy years was winding down. The military imperative that had been driving space technology in mid-century is giving way to an exploration imperative toward the end of the century. In that time, science and engineering have made human transportation in space feasible. A massive space technology has evolved from the Cold War and the enterprise of the Moon. It continues to evolve.

NASA's 90-day study of *Human Exploration of the Moon and Mars* presents a blueprint for establishing human presence on the Moon and Mars in the first two decades of the 21st century. As I have reported in this book, the technological means of moving out into the Solar System and of occupying new environments exist or are capable of being developed. Pioneering is not only a heritage of the national past but a prospect of the near future as well.

How near? The NASA Report cited five options or "reference approaches" for the return to the Moon and human exploration of Mars. The year 2001 is the earliest humans would return to the Moon and human habitation could begin there in 2002, the report proposed.

It said that a crew of four could land on Mars in 2011 for a 30-day stay and return to Earth the following year. Another option proposed that a crew of four lands in 2018 for a stay of up to 600 days and departs in 2020.

The report is a re-statement of a human exploration initiative first projected by NASA 20 years ago and updated since then. It has been imbued with new significance now, however, as a Presidential statement of national policy and intent.

"The imperative to explore is embedded in our history, our traditions and our national character," the report said. "Today, men and women have explored nearly every corner of the planet, even to the bleak center of remote Antarctica. Now, in the late 20th century and the early 21st, men and women are setting their sights on the Moon and Mars, as the exploration imperative propels us toward new discoveries."

NOTES

1. BEYOND TRANQUILITY

1. W. Von Braun, "Crossing the Last Frontier," *Colliers*, March 22, 1952.

2. U.S. Congress, Office of Technology Assessment, *Civilian Stations and the U.S. Future in Space* (Washington, D.C.: GPO, 1984).

3. Ibid.

4. James M. Grimwood, Lloyd S. Swenson, Jr., and Courtney G. Brooks, *Chariots for Apollo* (Houston: NASA, 1969).

5. Ibid.

6. Personal interview, September 18, 1975, as quoted in R. S. Lewis, *From Vinland to Mars* (New York: Quadrangle Books, 1976, 1978).

7. U.S. Congress, Office of Space Technology Assessment, p. 166.

8. Ibid., p. 167.

9. Ibid., pp. 168–169.

10. Ibid., pp. 169–170.

11. Ibid.

12. NASA Authorization Hearings, Subcommittee on Science, Technology, and Space, 96th Congress, 1st session, April–June 1979.

13. Associated Press, Washington, D.C., September 15, 1983.

2. BIRTH OF A STATION

1. *Aviation Week and Space Technology,* July 21 and September 23, 1986.
2. NASA Response to the NRC Report, April 1988, p. 50

3. THE GRAND DESIGN

1. NASA Response to the National Research Council Report, September 14, 1987.

4. TRANSITION

1. John M. Garvey, "Adaptation of Space Station Technology to Lunar Operations." Paper submitted to Lunar Base Symposium, NASA, Houston, April 4–8, 1988.
2. D. J. Weidman, W. M. Cirillo, and C. P. Llewellyn, "Study of the Use of the Space Station to Accommodate Lunar Base Missions." Paper presented at the Lunar Base Symposium, NASA, Houston, April 4–8, 1988.
3. *Aviation Week and Space Technology,* January 2, 1989.

5. THE NEW WORLD

1. K. A. Ehricke, "Lunar Industrialization and Settlement," Lunar Bases and Space Activities of the 21st Century Symposium, 1984.
2. S. W. Johnson and R. S. Leonard, "Evolution of Concepts for Lunar Bases." Paper presented at the Lunar Bases and Space Activities of the 21st Century Symposium, 1984.
3. Paul D. Lowman, Jr. "Lunar Bases and Post-Apollo Lunar Exploration." An Annotated Bibliography of Federally Funded American Studies, 1960–1982.
4. Ibid.
5. Ibid.
6. NASA Technical Memorandum, Ames Research Center (August 1973), vol. 1.
7. Third Lunar Science Conference, news conference, Houston, January 11, 1972.

6. A MANIFEST DESTINY

1. Luis W. Alvarez, Walter Alvarez, Frank Asaro, Helen V. Michel, "Extraterrestrial Cause for the Cretaceous-Tertiary Extinction." *Science* (June 6, 1980), vol. 208.
2. J. D. Blacic, "Structural Properties of Lunar Rock Materials under Anhydrous, Hard Vacuum Conditions." Symposium paper, 1984.

3. M. R. Sharpe, *Living in Space* (Garden City, NY: Doubleday, 1969).
4. Meek, T. T., Cocks, F. H., Vaniman, D. T., Wright, R. A. "Microwave Processing of Lunar Materials." Symposium paper, 1984.

7. POWER FROM THE MOON

1. Most of the helium 3 has diffused from the Earth and has been lost through the atmosphere to space. The residue is contained in underground natural gas. The underground U.S. strategic helium storage caverns contain 30 kilograms of helium 3. Another 200 kilograms might be obtained from the entire U.S. resource of natural gas.
2. Wittenberg, L. J., J. F. Santarius, and G. L. Kulcinski, *Lunar Source of Helium 3 for Commercial Fusion Power. Fusion Technology* (1986) vol 10, no. 167.
3. G. L. Kulcinski et al., Fusion Energy from the Moon for the 21st Century. Conference on Lunar Bases and Space activities in the 21st Century, Nasa, Houston, April 1988.
4. *Astrofuel for the 21st Century.* Pamphlet published by the College of Engineering, University of Wisconsin, 1988.
5. Kulcinski et al.
6. Wittenberg et al., *Lunar Source of Helium 3 for Commercial Fusion Power.*
7. Ibid.
8. Ibid.
9. Ibid.
10. Ibid.
11. I. N. Sviatoslavsky and M. Jacobs, Mobile Helium 3 Mining and Extraction System and Its Benefits Toward Lunar Base Self-sufficiency. Conference on Lunar Bases and Space Activities in the 21st Century.
12. Planetary Exploration Through the Year 2000, Part II. Solar System Exploration Committee, NASA Advisory Committee Report. Washington, D.C., 1986, p. 176.
13. Kulcinski et al.
14. E. M. Cameron, Helium Mining on the Moon, a Technical Report. Conference on Lunar Bases and Space Activities of the 21st Century.
15. Kulcinski et al.
16. J. F. Santarius, Lunar Helium 3 Fusion Propulsion and Space Development. Lunar Bases Conference June 1988.
17. Interview at La Jolla, California, October 16, 1982.
18. E. R. Podnieks and W. W. Roepke, "Mining for Lunar Base Support," Lunar Bases and Space Activities of the 21st Century, Houston, 1984.
19. NASA Technical Memorandum. "Feasibility of Mining Lunar Resources for Earth Use." Ames Research Center, August 1973.

20. Address to the Commonwealth Club of California, San Francisco, July 24, 1970.

21. Amanda Lee Moore, "Legal Responses." Lunar Bases and Space Activities in the 21st Century. Houston, 1984.

22. C. C. Joyner and H. H. Schmitt, "Extraterrestrial Law and Lunar Bases." Conference on Lunar Bases and Space Activities in the 21st Century. Houston 1984.

23. Parts 3 and 4, 96th Congress, 2d Sess., 1980.

24. J. I. Gabrynowicz, "Mankind Provisions Reconsidered: A New Beginning." Conference on Lunar Bases and Space Activities of the 21st Century. Houston, 1988.

25. "The Last Place on Earth" is the title of Roland Huntford's account of the race to the geographic South Pole by Amundsen and Scott in 1912. It was published by Atheneum, New York, 1985.

8. A NATURAL HISTORY OF MARS

1. Principally in Percival Lowell's *Mars as an Abode of Life* (New York: Macmillan, 1908).

2. *Journal of Geophysical Research* (September 30, 1977), vol. 82, no. 28, p. 4364.

3. M. H. Carr, ed. *Geology of the Terrestrial Planets* (Washington, D.C.: NASA, 1986).

4. Science briefing, NASA–Ames Research Center, October 8, 1985.

5. R. E. Arvidson, A. B. Binder, and K. L. Jones, "The Surface of Mars," *Scientific American*, March 1978.

6. Ibid.

7. Ibid.

8. M. H. Carr, "Mars: A Water Rich Planet." Symposium on Mars; evolution of its climate and atmosphere. Washington, D.C., July 17–19, 1987.

9. M. H. Carr, "Volatiles on Mars," *Icarus*, 68: 187.

10. M. H. Carr, ed. *Geology of the Terrestrial Planets*.

11. Ibid.

12. Symposium on Mars, 1986; *Science*, August 29, 1986.

13. Parker et al., "Geomorphic Evidence for Ancient Seas on Mars." Mars symposium.

14. H. J. Moore et al., "Surface Materials at the Landing Site," *Journal of Geophysical Research* (September 30, 1977), 82 (28): 4522.

15. Samuel Glasstone, *The Book of Mars* (Washington, D.C.: NASA, 1968).

16. B. D. Murray and M. D. Malin, "Polar Volatiles on Mars," *Science* (November 2, 1983) vol 182.

17. Arvidson et al.

18. B. M. French. "The Viking Discoveries" NASA Washington, D.C. 1980.

19. N. H. Horowitz, G. L. Hobby, and J. S. Hubbard, "The Carbon Assimilation Experiments," *Journal of Geophysical Research* (September 30, 1977) 82 (28): 4659.

20. K. Riemann et al., "The Search of Organic Substances . . .," *Journal of Geophysical Research* (September 30, 1977), vol. 82, no. 28, p. 4641.

21. N. H. Horowitz et al., p. 4661.

9. THE EMERGENCE OF MARS

1. K. Biemann et al., "The Search for Organic Substances and Inorganic Volatile Compounds on the Surface of Mars," *Journal of Geophysical Research* (September 30, 1977) 82 (28).

2. Planetary Exploration Through the Year 2000. Part II, Report by the Solar System Exploration Committee, NASA Advisory Council.

3. Samuel Glasstone, *The Book of Mars* (Washington, D.C.: NASA, 1968).

4. Ibid.

5. A. L. Bufkin and W. R. Johnson, II, "Conceptual Design Studies for Surface Infrastructure," *Manned Mars Missions,* NASA–Los Alamos, 1986.

10. CASTLES IN ORBIT

1. O. Hill and R. O. Wallace "Manned Mars Vehicles Design Requirements for Aerocapture." Manned Mars Missions, NASA–Los Alamos, 1986.

2. Manned Mars Mission summary report.

3. H. P. Davis, "A Manned Mars Mission Concept with Artificial Gravity." Manned Mars Missions.

4. A. L. Friedlander, J. C. Niehoff, and D. V. Byrnes, "Circulating Transportation Orbits Between Earth and Mars," AIAA/AAS Astrodynamics Conference, Williamsburg, Va., August 18, 1986.

5. Ibid.

6. J. Niehoff, "Manned Mars Mission Design." AIAA/ Planetary Society Joint Conference, Washington, D.C., July 1985.

7. Friedlander, et al.

8. The acronym is attributed to W. M. Hollister, "Castles in Space," *Aeronautica Acta,* January 1967.

9. S. J. Hoffman, A. L. Friedlander, and K. T. Nock, "Transportation Mode Performance Comparison for a Systained Manned Mars Base," AIAA/AAS Conference, Williamsburg, Va.

10. "Conceptual Sketch for an Aggressive Man in Space Program." Eagle Engineering, Report 85–109C, November 14, 1985.

11. S. J. Hoffman et al.

12. Eagle Report, 85–109C.

13. A. L. Bufkin and W. R. Jones, II. "Conceptual Design Studies for Surface Infrastructure," *Manned Mars Missions* (June 1986), vol. 1.

14. S. J. Hoffman et al.

INDEX

Aaron, John, 87
Adams, James H., Jr., 127
Adobe shelter, 130
Advanced Mars Base, 205
Advanced Solid Rocket Motor, 63
Advanced Technology Advisory Committee, 23
Advanced X-ray Astrophysics Facility, 80
Aerobrake, 211
Aerobrake technology, 83
Aerojet Solid Propulsion Co., 65
Aeroshell, 211
Agnew, Spiro T., 8
Air pressure variation, 170
Alba Patera, 168
Aldrin, Edwin E. (Buzz), Jr., 217
Alexandrov, Alexander, 96
Alvarez, Walter, 120
Ames Lunar Mining study, 157
Ames Research Center, 95, 106, 137, 220
Amundsen, Roald, 159
Antarctic analogy, 122
Antarctic treaty, 158-59
Apollo Applications Program, 4, 8, 103
Apollo 8, 6
Apollo 11, 1, 100
Apollo 12, 146
Apollo 13, 5
Apollo 15, 16, 151
Apollo 17, 5, 100, 155
Apollo 18, 19, 20, 5

Apollo Lunar Science Experiments, 103
Apollo Roving Vehicles, 102
Apollo Telescope Mount, 9
Army Ballistic Missile Agency, 3
Ariane, 71
Arnold, James R., 154
Armstrong, Neil, 100
Artificial gravity, 214, 215
Assembly sequence (sketches), 46
Astrofuel, 141
Astrotech Space Operations, 23
Atlantic Research Corp., 65
Atlantis, 61B, 26, 50
Atlas rocket, 16
Atmosphere of Mars, 163
Atomic Energy Commission, 152
Attached Payload Accommodation Equipment, 61
Averner, M. M., 133

Baseline configuration revised (space station), 38
Beggs, James M., 120
Bellingham, John, M.D., 97
Bernal, J. D. (on escape from Earth), 120
Biemann, Klaus, 184, 191
Biological extinctions, 120
Biomass Production Chamber, 134-36
Boeing Aerospace Co., 20, 48
Boland, Edward, U.S. Rep., 23, 68
Borman, Frank, 4
Briggs, Geoffrey A., 87, 89, 114
Briggs, Randall, 132
Buchanan, Paul, M.D., 93, 94
Budget turmoil, 70
Bugbee, Bruce G., 136
Burke, James D., 124
Byrd, Rear Adm. Richard E., 159

Cameron, Eugene N., 146
Carlucci, Frank C., 41
Carr, Michael H., 169, 172
Carter administration, 12
Cassini (Saturn) mission, 78, 87
CASTLE (in orbit), 217, 218, 219, 221
Catastrophic flooding, 175
Cave shelter, 131
Cernan, Eugene, 155
Challenger, 22, 25, 26, 74, 100, 189
Channels, Martian, 175-76
Chryse Planitia, 165, 170, 171
Circulating orbits, 203, 216
Cleave, Mary L., 27, 30
Clifford, Stephen, 177
Climate influences, 177
Climate variation, 169
Closed Environmental Control System, 22
Cocks, Franklin R., 133
Cold War, 2
Columbia, 62
Columbus (ESA), 22, 33, 56

Comet, Barnard, M.D., 98
Comet rendezvous, 78
Committee on Aeronautical and Space Sciences, Senate, 5
Common heritage principle, 157
Commonality doctrine, 107
Concrete shelter, 130
Conference on Lunar Bases, Washington, 114, 201
Conjunction class mission (Mars), 203, 211
Conrad, Charles Jr., 144
Construction flight schedule, space station, 58
Construction stretch-out, 68
Continent for Science, 159
Controlled Ecological Life Support System, 133, 136
Cooper, Leroy Gordon, 3
Copernicus crater, 167
Cosmic radiation dosage, 124
Cosmos 782, 936, 1129, 97
Cost estimates, space station, 43, 48
Cost reduction options, space station, 35
Costs compared, space station vs. lunar base, 111
Council of the ESA, 22

Darwin, Sir George B., 117
Das Mars Projekt, 161, 206
Defense Dept., 6
Deimos, 185
Delta rocket, 16
Deuterium-helium 3 fusion, 142
Discovery Mission 26, 83
Division of Solar System Exploration, 100
Donahue, Thomas M., 14
Douglas Aircraft Co., 3, 5
Down escalator orbit, 220
Dual keel design, 151
Duke, Michael B., 107, 112, 122
Duricrust, 180

Eagle Engineering Co., 100, 205, 219
Earth-Mars transfers, 196
Earth-Moon trade, 149
Earth-Moon transportation system, 110
Earth orbiting laboratory, 3
Earth return vehicle, 198
EASE, 26, 29, 31, 50
Ehricke, Krafft A., 100, 148
Electromagnetic accelerator, 152
Electromyostimulation, 45
Elysium Mons, 169
Emergency rescue, 42
Equatorial lander, Mars, 208
ESA, 22
ESA Platform, 40
Escalator orbit, 218
Expendable launcher, man-rated, 42
Extravehicular activity (EVA), 3

Fairchild Weston, 50
Fermi, Enrico, 41
First National Conference on the Peaceful Uses of Space, 119

Fletcher, James C., 6, 12, 22, 45, 68, 70, 76
Flight Telerobotic Servicer, 25, 38, 50
Ford Aerospace, 50
Fourth orbiter (OV 5), 36
Fourteenth Lunar-Planetary Conference, 107
Fra Secchi (Pietro Angelo), 160
Friedlander, Alan L., 217
Frosch, Robert A., 12
Fusion rockets, 147
Fusion Technology Institute, 141
Future National Space Objectives (report), 103

Gabrynowicz, Joanne Irene, 157
Galileo mission, 87
Garrett Airesearch, 50
Garriott, Owen, 93, 94
Gazenko, Oleg, M.D., 98
Gelfand, Eric M., 106
Gemini program, 26
General Dynamics Co., 50
General Electric Co., 48
Geostationary orbit, 2
Gilruth, Robert R., 5, 7
Global positioning satellite, 35
Goddard Space Flight Center, 17, 20, 201
Graham, William R., 41
Gravity gradient stabilization, 44
Green, Bill, U.S. Rep., 23
Greenhouse agriculture (Mars), 220
Grumman Co., 25, 50

Habitation module, 40, 54, 56
Hale, Edward Everett, 2
Hamilton Standard, 50
Harris Corp., 50
Helium 3, 89, 101, 142
Helium 3 fusion reaction, 140
Helium 3 mining, 147
Hercules, Inc., 65
Hill, Walter, 136
Hodge, John D., 15, 24
Hoffman, Stephen J., 121
Hohman, Walter, 203
Honeywell, 50
Horizon project, 3
Horz, Friedrich, 131

IBM, 50
Ilyn, Evgeny, M.D., 99
Industrial Space Facility, 23, 25, 70
Intelsat, 156
International Telecommunications Union, 157
Intraspecific Competition, 2, 100
Irwin, James B., 151

Jacobs, Mark, 143
Japanese Experiment Module, 22, 34, 40, 41, 57
Johnson Space Center, 13, 18, 22
Joint Committee on Atomic Energy, 210
Joyner, Christopher C., 156

Kaplicky, Jan, 128
Keaton, Paul W., 106, 139
KSC Biomass Production Center, 12, 39, 138
Keyworth, George, 14
Khalile, Nader E., 130
KIWI, reactor, 154
Kizim, Leonid, 32
Klein, Harold P., 133, 182, 185
Knott, William M., 136
KREEP (lunar), 116, 154
Kulcinski, Gerald L., 141

Labelled Release Experiment, 183, 191
Laboratory incubators (Viking), 166
Laboratory, U.S. (space station), 40
Land, Peter, 27
Langley Research Center, 3, 6, 21, 35, 72
Laser wind sounder, 79
Launch vehicle, heavy lift, 41
Lava tube shelter, 131, 132
Lewis Research Center, 17, 21, 209
L-1 Libration point, 221
Lifeboat, 35
Life support systems, 102
Lin, T. D., 129
Lockheed Missiles and Space Co., 50
Long March rocket, 71
Los Alamos conference, 1984, 112
Low, George M., 5, 210
Lowman, Paul D., Jr., 120
Luna spacecraft, 193, 198, 200
Lunar aluminum, 12
Lunar assets, 140
Lunar base, 107
Lunar base construction, 125
Lunar base cost study, 139
Lunar base design, 124-28
Lunar base development, 101, 102
Lunar base plan, 121
Lunar Bases & Space Activities of the 21st Century, 119
Lunar concrete, 117
Lunar drilling, 151
Lunar economic development, 122
Lunar environment assets, 123
Lunar exploration systems, Apollo, 102
Lunar Far Side observatory, 109
Lunar Geoscience Orbiter, 100, 102, 114, 123
Lunar greenhouse, 105
Lunar ice, 114
Lunar industry, 12
Lunar magnetism, 117
Lunar mining, 151
Lunar module, 121
Lunar nuclear power systems, 152
Lunar Orbiter, 89, 123
Lunar orbit station, 104
Lunar oxygen production, 132
Lunar Planum, 176
Lunar polar base, 123
Lunar polar science, 123

Lunar processing plant, 155
Lunar Receiving Laboratory, 101, 105, 199
Lunar resources, 113
Lunar science working group, 118
Lunar solar power satellites, 105
Lunar upper mantle, 116
Lunar volcanism, 116
Lunokhod, 78, 199

MacElroy, R. D., 133
McDonald Douglas Astronautics, 11, 20, 48
Magellan, probe, 78, 87
Manarov, Musa, 96
Manifest Destiny, 121
Manned lunar laboratory, 107
Manned Mars Mission study, 214
Manned mission module, 219
Man tended station, 22, 31, 59
Manned Orbital Research Laboratory, 19
Manned Orbital Systems Concept, 11
Mare Tranquillitatis, 1
Mariner 4, 8, 162, 188
Mariner 6, 7, 8, 162, 172
Mariner 9, 165, 168, 193
Maiorski, Boris, 158
Mark, Hans, 112
Mars, 1
Mars ascent vehicles, 198
Mars base, 211
Mars Departure Vehicle, 211
Mars Excursion Vehicle, 208
Mars Exploration Science Advisory Group, 197
Mars fly-by mission, 202
Mars ice age, 169
Mars Mission Working Group, 215
Mars Observer, 78, 81, 191
Mars (flight) opportunity, 203
Mars Working Group report, 201-2
Mars Sample Return/Rover, 78, 91, 195
Mars transport, 209, 213
Marshall Space Flight Center, 11, 13, 20
Mars water resources, 204
Martin Marietta Aerospace, 20, 25
Mathews, Charles W., 7
Mauna Loa (compared with Mars volcanoes), 168
Meek, Thomas T., 133
Mendell, Wendell W., 107, 122
Mercury Atlas 9, 3
Microgravity, 3
Microgravity conference, 96
Microgravity research, 72, 74
Miller, James C II, 41
Mining on the Moon, 142
Mining payback, 144
Minimal Mars Base, 204
Mir space station, 25, 32, 45, 47, 82, 96
Mission Peculiar Equipment Support Structure, 28
Mission to Planet Earth, 77
Mobile Lunar Mining Machine, 143
Mobile Servicer, 22, 24, 33, 40, 59, 60

MOLAB, 103
Montmorillonite (clay), 182
Moon: as an industrial annex, 113; as a planet, 101; as a Rosetta stone, 118
Moon base facilities, 110
Moonlab study, 106
Moon treaties, 156-57
Moore, Amanda Lee, 156
Morton Thiokol Co., 65
Mueller, George F., 5, 9
Mulrooney, Brian, 24
Muscle atrophy in microgravity, 92

NASA, 1
NASA Advisory Committee Report on Manned Space Flight, 103; Dept. of Exploration, 190; Division of Solar System Exploration, 76, 87, 102; Lunar Exploration Conference, 102; Office of Advanced Research and Technology, 102; Office of Space Station, 87; Site Evaluation Board, 66; Space Nuclear Propulsion Office, 152
NASA/AIR FORCE heavy lift vehicle, 81
National Commission on Space, 74, 91, 190, 200
National Research Council Review, 41, 42
NERVA (nuclear rocket engine), 104, 153, 209
Neudecker, Joseph W., 132
Next Logical Step, 12
Niehoff, John C., 121, 217
Nix Olympica (Olympus Mons), 167, 170
Nixon, David, 128
Nixon (White House), 6
Node, 48, 57
Nuclear fusion rocket, 153
Nuclear thermal propulsion, 210

Oberth, Hermann, 3
O'Connor, Byron D., 27
Odom, James, 70
Office of Advanced Research & Technology (NASA), 4
Office of Manned Space Flight (NASA), 4
Office of Space Science and Applications, 4
Office of Technology Assessment, 3, 4, 6, 7, 13
Opposition class mission, Mars, 203
Orbital Maneuvering & Orbital Transfer vehicles, 15, 20, 47, 48, 70, 84, 107, 121, 200
Outer Space Treaty, 156
Outlook for Space, report, 12, 105

Paine, Thomas O., 6, 74, 75
Palapa B., 26
Panama Canal, 66
Parker, Timothy J., 177
Pathfinder Program, 70
Penguin suit, 93
Periodic Space Station, 219
Permafrost, 172
Phase I (space station), 46
Phase II (space station), 47, 48, 59
Phobos, 91, 163, 184, 206
Photovoltaic power arrays, 49
Pioneering the Space Frontier, report, 76

Planetary Society, 91, 210
Pleistocene Epoch, 176
Podnieks, E. R. (Bureau of Mines), 151
Polar glaciers (Mars), 170, 172, 208
Polar orbiting platform, U.S., 32, 40
Post-Challenger changes, 77, 83
Post lunar space goals, 189
Power Tower (space station), 21, 22, 32
Precision Pointing Mount, 61
Presidential Commission on the Challenger Accident, 63
President's Science Advisory Commission, 6
Pressurized carriers, 58
Private investment in space, 150
Project Apollo (review), 206
Project Mercury, 2, 3
Proton rocket, 71
Proxmire, William, Sen., 23, 68

Radiation exposure standards, 127
Radiation shelter, 220
Radioisotope Thermal Electric Generator, 152
RCA, 50
Reagan space policy, 2, 22, 68, 70, 72, 100, 189
Re-boost System, 42
Remote Manipulator System, 20, 22, 23
Ride, Sally, astronaut, STS 7, 61
Ride Report, 76, 77, 81, 82
River channels, Mars, 174
Remote Manipulator System, 30
Roberts, Barney B., 122
Rockedyne Div., 21, 48
Rockwell, International, 20
Roepke, W. W. (Bureau of Mines), 151
Rogers Commission, 76
Romanenko, Yuri, 96
Rover vehicle, 192
Ross, Jerry L., 26, 50
Rowley, John C., 132
Runcorn, Keith, 117
Ruppe, Harry O., 208

S4B (Saturn 5 stage), 5, 9
Sacco, Albert, Jr., 132
Safe haven, 35, 220
Sagan, Carl, 98, 186
Salisbury, Frank B., 136
Salyut, 9
Sand dunes, Mars and Earth compared, 189, 190
Santarius, John F., 147
Saturn 5 rocket, 1, 26
Saturn 5-Apollo, 3, 7, 121
Seamans, Robert C., Jr., 3, 41
Seabed mining rights, 157
Seaborg, Glen T., 153
Schiaparelli, Giovanni, 160, 162
Schmitt, Harrison H., 4, 20, 155, 156
Schroter, Johannes, 160
Science Applications International Corp., 108, 121, 193, 210
Scott, David R., 151
Scotch verdict on Mars, 187

Search for life, 182
Sellers, Wallace O., 139
Senate Committee on Aeronautical & Space Sciences, 157, 210
Senate Subcommittee on Science, Technology and Space, 15, 24
Senior Interagency Group for Space, 15
Shackleton, Sir Ernest, 122, 159
Shapiro, Maurice M., 127
Sharpe, Terry H., 104
Shaw, Brewster A., Jr., 27
Shuttle deployable mast, 90
Shuttle main engine system, 210
Shuttle traffic model, 1988–94, 55
Single truss space station design, 32
Skylab, 8, 9, 19, 26, 31, 44, 56; lab damage, 10
SNAP 8 (System for Nuclear Auxiliary Power), 154
SNAP 19, 153
SNAP 27 (jettisoned), 154
Solar Maximum Mission Satellite, 26
Solar Dynamic Power System, 25, 49
Solar luminosity change, 191
Solar System Exploration Committee, 14, 145, 189, 191
Solar Thermal Power System, 53
Solar wind fallout, 116, 140
Solid rocket motor plan, 63
Solovyev, Vladimir, 32
Soviet missions (Mars), 181
Soviet space station, 26
Space adaptation syndrome, 92
Space Industries, Inc., 23
Space Science Board conference, 102
Spacelab, 8, 18, 33, 62
Space Operations Center, 13
Space Station construction, 8, 54
Space Station Task Group, 1, 2, 6, 8, 14, 15, 101, 209
Space tug, 104
Spring, Sherwood C., 26, 29, 50
State Department, 6
State of the Union message, 1984, 17
Stofan, Andrew, 36, 70
Station revised budget, 35
Stennis Space Center (National Space Technology Laboratory), 66
Stepping Stones to Mars, 209
Strategies for a lunar base, 122
Sub-dry ice reservoir, Mars, 180
Subterrene tunneling system, 133
Sviatkoslavsky, Igor N., 143
Swing-by of Mars, 211
Synthetic Aperture Radar, 77
Systems Requirements Review, 32, 35

Taurus-Littrow, 100
Taxi (space vehicle), 217
Tempel 2, 78
Teledyne Brown Engineering, 50
Telerobotic Servicer, 33, 60
Terminal cataclysm, 115
Tharsis Ridge, 168

Third party liability (in space activities), 72
Titan rocket, 32, 165, 197
Titan, moon of Saturn, 77
Titov, Vladimir, 96
Toulmin, Priestley III, 186
Tracking and Data Relay Satellite, 17, 35, 83, 85, 86, 88
Truly, Adm. Richard H., 87
Truman, Harry, 4
TRW, 21, 50
Tsiolkovsky, Konstantin, 2, 97, 222
TVA, Yellow Creek plant, 66
Tyuratam spaceport, 47

UN Conference on Law of the Sea, 157
U.S. Bureau of Mines testing, 151
United Technologies Corp., 65
Up Escalator orbit, 222
Ushakov, Arkady, 96
Utopia Planitia, 166, 170

Valles Marineris, 162, 164
Van Allen Zone, 209
Vandenberg Air Force Base, 4, 66

Van den Berg, Lodwijk, 16
Vaniman, David T., 133
Venus, 189, 210
Viking, 165, 188, 191, 193
VISIT circulating orbit, 216
Von Tiesenhausen, Georg, 104
Von Braun, Wernher, 3, 7, 9, 120, 161
Vostok, 9

Wilkes, Lt. Charles, 158
Ward, William R., 173
Webb, James E., 4, 6
Weight constraint, 26
Wells, H. G., 161
Westar 6, 26
Westinghouse Electric Co., 23
Woods Hole Oceanographic Institution, 12
Work packages, space station, 20, 21, 48, 49, 133

Yield from helium 3-deuterium reaction, 142
Young, John W., 151

Zero gravity (see microgravity)